Echoes of the Old Country

Echoes of the *Old Country*

Growing Up German-Russian on the Northern Plains

by Jessica Clark

NDSU NORTH DAKOTA STATE UNIVERSITY PRESS

Fargo, North Dakota

North Dakota State University Press
Dept. 2576, P.O. Box 6050, Fargo, ND, 58108-6050
www.ndsupress.org

Echoes of the Old Country: Growing Up German-Russian on the Northern Plains
By Jessica Clark

First Edition
First Printing

LCCN: on order at time of publication
ISBN: 978-1-946163-64-6 (hardcover)
ISBN: 978-1-946163-79-0 (paperback)

Cover design by Robert Angelo
Interior design by Deb Tanner

The publication of *Echoes of the Old Country: Growing Up German-Russian on the Northern Plains* is made possible by the generous support of donors to the NDSU Press Fund and the NDSU Press Endowed Fund, and by other contributors to NDSU Press.

Kimberly Wallin, Director
Suzzanne Kelley, Publisher
Kyle Vanderburg, Assistant Acquisitions Editor
Emma Borah, Graduate Research Assistant in Publishing
Kaelyn McHugh and Ellie Keppel, Editorial Interns
Book Team for *Echoes of the Old Country: Growing Up German–Russian on the Northern Plains*: Ella Hlucny, Monika Sauer, Brandon Stariha, Breanna Walburn

Printed in the United States of America

Publisher's Cataloging-in-Publication
(Provided by Cassidy Cataloguing Services, Inc.)

Names: Clark, Jessica (Jessica N.), author.
Title: Echoes of the old country : growing up German-Russian on the Northern Plains / by Jessica Clark.
Description: First edition. | Fargo, North Dakota : North Dakota State University Press, [2025] | Include bibliographical references and index.
Identifiers: ISBN: 9781946163646 (hardcover) | 9781946163790 (paperback)
Subjects: LCSH: German Americans--Great Plains--History--20th century. | German Americans--Great Plains--Social life and customs. | Russian Germans--Great Plains--History--20th century. | Germans--Russia--History. | Collective memory--Great Plains. | Coming of age--Great Plains. | Farm life--Great Plains. | Great Plains--History--20th century. | LCGFT: Oral histories. | BISAC: HISTORY / United States / State & Local / Midwest (IA, IL, IN, KS, MI, MN, MO, ND, NE, OH, SD, WI) | SOCIAL SCIENCE / Emigration & Immigration. | HISTORY / United States / 20th Century.
Classification: LCC: F596.3.G3 C53 2025 | DDC: 978.00431--dc23

Dedication

This book is dedicated to my mentor and friend, Thomas D. Isern,
whose guidance, support, and unwavering belief in me
shaped not only this work but also the leader I have become.
His dedication to access, opportunity, and excellence has inspired me
to carry those values forward in all that I do.

It is also dedicated to the narrators of the Dakota Memories
Oral History Project, whose voices breathe life into these pages.
To those who have passed: your stories remain, and I hope
your families see you reflected in this work.

Table of Contents

Preface

This book is devoted to the memories of the second- and third-generation Germans from Russia. It is not an ethnic or generational comparison. Rather, this work focuses on the voices and memories of insiders. *Echoes of the Old Country* is a study of their childhood memories and ethnic identity, on their terms and in their words. In essence, it is an analysis through narrative, a writing style employed by the renowned oral historian Alessandro Portelli, who observes that an oral history book "is made of many voices. The narrators are its co-authors."[1]

Using more than 200 oral histories in *The Order Has Been Carried Out* and more than 150 in *They Say in Harlan County*, Portelli has mastered a technique of literary analysis he dubs as narrative voice. According to him, his role—as the historian, as the author—is to construct "a coherent discourse [to make his] . . . point out of the voices of many others."[2] He compares his role to that of a quilter, taking bits and pieces, fragments, if you will, and piecing them together to reconstruct a larger product. Yet, this larger product is comprised of original, unaltered pieces. He also compares his work to that of a montage in cinema or a sampling in music. Regardless, in his works of history, from Nazi massacres in Rome to coal miners in the United States, Portelli lets the voices speak for themselves.[3]

From 2005 to 2010, nearly two hundred German-Russians of the second and third generation agreed to have their voices recorded by oral historians associated with the Dakota Memories Oral History Project. Concerned foremost

---

[1] Alessandro Portelli, *The Order Has Been Carried Out: History, Memory, and Meaning of a Nazi Massacre in Rome* (New York: Palgrave Macmillian, 2003); Portelli, *They Say in Harlan County: An Oral History* (Oxford University, 2010).

[2] Portelli, *They Say in Harlan County*, 11.

[3] Portelli, *The Order Has Been Carried Out.*

with collecting the voices and memories of the narrators, these oral histories were recorded and logged but not transcribed. Indeed, less than 5 percent of the oral histories have been transcribed to date. The aim of this project was to record and to preserve the narrations in their original form, which include voice fluctuations and body language; thus, all the oral histories were video recorded. The process was as much about the aural as the oral.

Participants of this project were well aware that their oral histories were being recorded for the purpose of historical research. They each signed a consent form granting the Germans from Russia Heritage Collection (GHRC), located in the North Dakota State University Library, permission to reproduce and make available their oral histories. In addition, each participant was aware of my intention to incorporate their voices into a work on childhood memories of German-Russian descendants. Narrators of Dakota Memories project willingly shared their memories and their voices with the understanding that their stories would be told.

That said, this book is my interpretation of their memories, their voices, and their stories. Using Portelli's approach, this work is a

> multi-voiced narrative, a montage of fragments of varying length . . . [because] oral history is not just the collection of stories but also their interpretation and representation. The interpretation begins with the selection of sources, continues in the researcher's active role in the interview, and culminates in the open comments of the authorial voice and the meanings implicit in editing and montage.[4]

Employing his technique of narrative analysis, *Echoes of the Old Country* incorporates the many voices of the Dakota Memories participants. By piecing these narratives together with the existing historiography, I have revealed layers of complexity previously unseen. The result is a new German-Russian story, as told by the voices of the Dakota Memories Oral History Project.

Below is a list of all the participants whose voices were used to construct this new narrative. Included in the list is basic data for each narrator, formatted with each narrator's name, nicknames and middle names if known, and maiden names (in parentheses) if known. The data also includes the year and place of birth (if known) and the year of death. The oral historians' names are provided, too, along with the place and date of the interviews. The GRHC houses the

[4] Portelli, *The Order Has Been Carried Out*, 17–18.

original videotaped oral histories and archival materials (diaries, photographs, memories, family documents, etc.) in the Dakota Memories Oral History Project digital archives, which are located at the Digital Horizons website.[5]

Axtmann, Nicholas F., b. 1919 in Tuscorora Township, North Dakota; d. 2009 in Harvey, North Dakota. Interview by Jessica Clark in Rugby, North Dakota, on 7 July 2006.

Backfish, Hilda Melita (Lindemann), b. 1928 in Golden Valley, North Dakota; d. 2016 in Beulah, North Dakota. Interview by Cassie Ptacek in Beulah, North Dakota, on 9 June 2008.

Bauman, Albert, b. 1909 in Mercer County, North Dakota; d. 2009 in Hazen, North Dakota. Interview by Cassie Ptacek in Golden Valley, North Dakota, on 15 June 2008.

Bauman, Clarence, b. 1923 in Java, South Dakota; d. 2023 in Bismarck, North Dakota. Interview by Jessica Clark in Bismarck, North Dakota, on 22 July 2007.

Bauman, Marilyn Jeanette (Sauer), b. 1932 in Java, South Dakota; d. 2014 in Bismarck, North Dakota. Interview by Jessica Clark in Bismarck, North Dakota, on 22 July 2007.

Becker, Delores (Kuhn), b. 1933 in McIntosh County, North Dakota. Interview by Andrea Mott in Napoleon, North Dakota, on 3 June 2009.

Becker, John G., b. 1929 in Bismarck, North Dakota; d. 2022 in Bismarck, North Dakota. Interview by Andrea Mott in Napoleon, North Dakota, on 3 June 2009.

Bender, Arthur R., b. 1922 in Wishek, North Dakota; d. 2010 in Wishek, North Dakota. Interview by Sarah (Lacher) Crary in Wishek, North Dakota, on 10 August 2006.

Bendewald, Reinhold, b. 1924 in Ashley, North Dakota; d. 2019 in Ashley, North Dakota. Interview by Cassie Ptacek in Ashley, North Dakota, on 21 May 2008.

Berger, George L., b. 1927 in Dickinson, North Dakota; d. 2015 in Glendive, Montana. Interview by Andrea Mott in Dickinson, North Dakota, on 25 June 2009.

Berger, Pauline "Polly" (Wock), b. 1929 in Stark County, North Dakota; d. 2018 in Dickinson, North Dakota. Interview by Cassie Ptacek in Dickinson, North Dakota, on 4 August 2007.

Berger, Rosemary (Fisher), b. 1929 in Dickinson, North Dakota. Interview by Andrea Mott in Dickinson, North Dakota, on 25 June 2009.

---

[5] The Dakota Memories Oral History Project (GRHC), Digital Horizons: Life on the Northern Plains, https://cdm16921.contentdm.oclc.org/.

Bertsch, Emmanuel, b. 1927 in Scott, Saskatchewan. Interview by Jessica Clark in Saskatoon, Saskatchewan, on 21 July 2006.

Bitz, Margaret, b. 1938 in Logan County, North Dakota; d. 2022 in Napoleon, North Dakota. Interview by Andrea Mott in Fargo, North Dakota, on 5 May 2009.

Black, Elsie Ellen (Brick), b. 1924 in Devils Lake, North Dakota; d. 2016 in Minot, North Dakota. Interview by Cassie Ptacek in Knox, North Dakota, on 22 June 2007.

Black, Joseph John, b. 1924 in Berwick, North Dakota; d. 2015 in Minot, North Dakota. Interview by Jessica Clark in Knox, North Dakota, on 6 July 2006.

Boechler, Raymond, b. 1925 in Karlsruhe, North Dakota; d. 2016. Interview by Jessica Clark in Saskatoon, Saskatchewan, on 21 July 2006.

Boechler, Wilfred Jacob, b. 1926 in Allan, Saskatchewan; d. 2020 in Wakaw, Saskatchewan. Interview by Jessica Clark in Allan, Saskatchewan, on 23 July 2006.

Boschee, Adam Gordon, b. 1945 in Kulm, North Dakota. Interview by Jessica Clark in West Fargo, North Dakota, on 17 April 2005.

Brossart, Valentine F., b. 1926 in Rugby, North Dakota; d. 2008 in Rugby, North Dakota. Interview by Jessica Clark in Rugby, North Dakota on 5 July 2006.

Brosz, Betty (Krueger), b. 1932 in Fredonia, North Dakota. Interview by Eric Mischnick in Fredonia, North Dakota, on 29 May 2008.

Buck, Alice Ruth (Miller), b. 1932 in Streeter, North Dakota. Interview by Jessica Clark in Streeter, North Dakota, on 23 May and 24 June 2005.

Buck, Edna Lu Ella (Weisz), b. 1923 in Lehr, North Dakota; d. 2010. Interview by Cassie Ptacek in Fargo, North Dakota, on 3 May 2008.

Buck, LeRoy, b. 1930 in Streeter, North Dakota. Interview by Jessica Clark in Streeter, North Dakota, on 23 May and 24 June 2005.

Buerkley, Alice (Woehl), b. 1930 in Fredonia, North Dakota; d. 2021 in Jamestown, North Dakota. Interview by Cassie Ptacek in Kulm, North Dakota, on 2 June 2008.

Degenstein, Leo James, b. 1928 in Chicago, Illinois; d. 2013 in Minot, North Dakota. Interview by Jessica Clark in Rugby, North Dakota, on 9 July 2006.

Degenstein, Magdalena "Maggie" (Kuntz), b. 1935 in Orrin, North Dakota. Interview by Jessica Clark in Rugby, North Dakota, on 9 July 2006.

Dewald, Edwin "Emil," b. 1933 in Napoleon, North Dakota; d. 2022 in Fargo, North Dakota. Interview by Jessica Clark in Fargo, North Dakota, on 26 April 2005; interview by Cassie Ptacek and Jessica Clark in Fargo, North Dakota, on 14 May 2007.

Dobler, Stanley Graven, b. 1930 in McIntosh County, North Dakota; d. 2023 in Ashley, North Dakota. Interview by Sarah (Lacher) Crary in Ashley, North Dakota, on 9 August 2006.

Dockter, Howard R., b. 1927 in Streeter, North Dakota. Interview by Eric Mischnick in Gackle, North Dakota, on 3 June 2008.

Doll, Anna (Sturn), b. 1928 in Crown Butte, North Dakota. Interview by Jessica Clark in Mandan, North Dakota, on 19 July 2007.

Drefs, Esther (Haerter), b. 1917 in Venstil, Montana; d. 2014 in Ashley, North Dakota. Interview by Jessica Clark in Ashley, North Dakota, on 28 June 2005.

Ebach, Mary, b. unknown in Rugby, North Dakota. Interview by Jessica Clark in Rugby, North Dakota, on 3 July 2006.

Eckroth, Fr. Leonard A., b. 1932 in Mandan, North Dakota. Interview by Sarah (Lacher) Crary in Strasburg, North Dakota, on 28 May 2006.

Eissinger, Theresa (Meier), b. 1929 in Napoleon, North Dakota. Interview by Andrea Mott in Napoleon, North Dakota, on 1 June 2009.

Engelhart, Florence (Mitzel), b. 1937 in Strasburg, North Dakota; d. 2017 in Ashley, North Dakota. Interview by Sarah (Lacher) Crary in Venturia, North Dakota, on 30 May 2006.

Engelhart, James, b. 1936 in Venturia, North Dakota; d. 2024 in Ashley, North Dakota. Interview by Sarah (Lacher) Crary in Venturia, North Dakota, on 29 May 2006.

Erbele, Raymond, b. 1945 in Jamestown, North Dakota. Interview by Andrea Mott in Streeter, North Dakota, on 10 June 2009.

Erbele, Robert, b. 1952 in Jamestown, North Dakota. Interview by Sarah (Lacher) Crary in Lehr, North Dakota, on 8 August 2006.

Fercho, Calvin K., b. 1925 in Bismarck, North Dakota. Interview by Jessica Clark in Fargo, North Dakota, on 27 April 2005.

Ferner, Ben, b. 1928 in Katharinental, Saskatchewan. Interview by Jessica Clark in Regina, Saskatchewan, on 29 July 2006.

Ferner, Clara (Ell), b. 1933 in Kronau, Saskatchewan. Interview by Jessica Clark in Regina, Saskatchewan, on 29 July 2006.

Fichter, Maryann (Doll), b. 1914 in Crown Butte, North Dakota; d. 2012 in Dickinson, North Dakota. Interview by Andrea Mott in Dickinson, North Dakota, on 27 June 2009.

Fiechtner, Henrietta (Gall), b. 1939 in Wishek, North Dakota; d. 2022 in Fargo, North Dakota. Interview by Cassie Ptacek in Fargo, North Dakota, on 13 June 2007.

Fiechtner, Marvin G., b. 1926 in Wishek, North Dakota; d. 2012 in Bismarck, North Dakota. Interview by Jessica Clark in Wishek, North Dakota, on 14 July 2005.

Fischer, ElRoy, b. 1923 in Gackle, North Dakota; d. 2017 in Fargo, North Dakota. Interview by Jessica Clark in Gackle, North Dakota, on 9 June 2005.

Fischer, Ruth (Dockter), b. 1925 in Streeter, North Dakota; d. 2017 in Wishek, North Dakota Interview by Jessica Clark in Gackle, North Dakota, on 9 June 2005.

Flegel, Marjorie (Miller), b. 1934 in Dickey County, North Dakota; d. 2023 in Wishek, North Dakota. Interview by Cassie Ptacek in Kulm, North Dakota, on 27 May 2008.

Frank, Emil, b. 1921 in Dodge, North Dakota, d. 2009 in Bismarck, North Dakota. Interview by Cassie Ptacek in Beulah, North Dakota, on 11 June 2008.

Fregien, Audrey A. (Lemke), b. 1922 in Wirch, North Dakota. Interview by Eric Mischnick in Kulm, North Dakota, on 28 May 2008.

Gab, Robert R., b. 1926 in Hillsview, South Dakota, d. 2019 in Eureka, South Dakota. Interview by Cassie Ptacek in Eureka, South Dakota, on 15 May 2008.

Gebhardt, Ervin, b. 1928 in McPherson County, South Dakota; d. 2021 in Eureka, South Dakota. Interview by Eric Mischnick in Eureka, South Dakota, on 20 May 2008.

Gebhardt, Lillian Elsa (Oster), b. 1927 in Campbell County, South Dakota; d. 2019 in Eureka, South Dakota. Interview by Cassie Ptacek in Eureka, South Dakota, on 20 May 2008

Geiszler, Erwin, b. 1919 in Logan County, North Dakota; d. 2012 in Bismarck, North Dakota. Interview by Andrea Mott in Napoleon, North Dakota, on 4 June 2009.

Graf, Alvin E., b. 1915 in Medina, North Dakota; d. 2010 in Bismarck, North Dakota. Interview by Cassie Ptacek in Jamestown, North Dakota, on 18 June 2008.

Gross, John J., b. 1924 in Napoleon, North Dakota; d. 2025 in Napoleon, North Dakota. Interview by Andrea Mott in Napoleon, North Dakota, 8 June 2009.

Gross, Margaret (Schaffer), b. 1927 in Napoleon, North Dakota; d. 2019 in Napoleon, North Dakota. Interview by Andrea Mott in Napoleon, North Dakota, on 8 June 2009.

Gross, Mathias "Math" A., b. 1921 in Logan County, North Dakota; d. 2013 in Bismarck, North Dakota. Interview by Jessica Clark in Bismarck, North Dakota, on 20 July 2007.

Hartmann, Marvin L., b. 1932 in Delmont, South Dakota. Interview by Jessica Clark in Fargo, North Dakota, on 26 April 2005.

Hauck, Magdalena (Staudinger), b. 1924 in Dickinson, North Dakota; d. 2017 in Richardton, North Dakota Interview by Cassie Ptacek in Richardton, North Dakota, on 9 August 2007.

Hauck, Melfrieda "Millie" Marie (Doll), b. 1931 in Blue Grass, North Dakota. Interview by Andrea Mott in Richardton, North Dakota, on 28 June 2009.

Hauck, Paul A., b. 1921 in Richardton, North Dakota; d. 2014 in Richardton, North Dakota. Interview by Cassie Ptacek in Richardton, North Dakota, on 9 August 2007.

Haupt, Wilma (Beck), b. 1925 in Eureka, South Dakota; d. 2015 in Eureka, South Dakota. Interview by Cassie Ptacek in Eureka, South Dakota, on 16 May 2008.

Hecker, Victoria (Schaff), b. 1924 in Belfield, North Dakota; d. 2015 in Dickinson, North Dakota. Interview by Andrea Mott in Dickinson, North Dakota, on 20 June 2009.

Helbling, Jacob A., b. 1921 in Morton County, North Dakota; d. 2013 in Bismarck, North Dakota. Interview by Jessica Clark in Bismarck, North Dakota, on 24 July 2007.

Helbling, Stanley, b. 1916 in Mandan, North Dakota; d. 2012 in Mandan, North Dakota. Interview by Jessica Clark in Bismarck, North Dakota, on 24 July 2007.

Helfrich, Louis, b. 1927 in Glen Ullin, North Dakota; d. 2018 in Morris, Minnesota. Interview by Cassie Ptacek in Dickinson, North Dakota, on 7 August 2007.

Hellmuth, Laura (Schulz), b. 1919 in Kintyre, North Dakota; d. 2023 in Kintyre, North Dakota. Interview by Andrea Mott in Napoleon, North Dakota, on 2 June 2009.

Hellmuth, William, b. 1912 in Burnstad, North Dakota; d. 2011. Interview by Andrea Mott in Napoleon, North Dakota, on 2 June 2009.

Herauf, Leo Peter, b. 1931 in Davin, Saskatchewan. Interview by Jessica Clark in Regina, Saskatchewan, on 30 July 2006.

Herman, Herbert, b. 1915 in Gackle, North Dakota; d. 2018 in San Jose, California. Interview by Jessica Clark in Gackle, North Dakota, on 25 May and 9 June 2005.

Herr, Richard, b. 1929; d. 2019. Interview by Sarah (Lacher) Crary in Wishek and Lehr, North Dakota, on 12 August 2006.

Heyne, August, b. 1913 in Grant County, North Dakota; d. 2010 in Bismarck, North Dakota. Interview by Jessica Clark in Bismarck, North Dakota, on 21 July 2007.

Heyne, Martha M. (Brosy), b. 1921 in Wishek, North Dakota; d. 2012 in Wishek, North Dakota. Interview by Jessica Clark in Wishek, North Dakota, on 14 July 2005.

Hoff, Leona (Kuhn), b. 1929 in Dunn County, North Dakota; d. 2017 in Richardton, North Dakota. Interview by Andrea Mott in Richardton, North Dakota, on 24 June 2009.

Hoffart, Anne (Frehlich), b. 1927 in Scott, Saskatchewan; d. 2021 in Unity, Saskatchewan. Interview by Jessica Clark in Unity, Saskatchewan, on 16 July 2006.

Hoffart, Frank, b. 1925 in Kerrobert, Saskatchewan; d. 2013 in Unity, Saskatchewan. Interview by Jessica Clark in Unity, Saskatchewan, on 16 July 2006.

Huber, Donald T., b. 1929 in Hazen, North Dakota; d. 2017 in Beulah, North Dakota. Interview by Cassie Ptacek in Hazen, North Dakota on 12 June 2008.

Hutmacher, Alexander, b. 1950 in Dickinson, North Dakota. Interview by Andrea Mott in Dickinson, North Dakota, on 22 June 2009.

Jahner, Michael, b. 1935 in Kerrobert, Saskatchewan; d. 2010. Interview by Jessica Clark in Tramping Lake, Saskatchewan, on 14 July 2006.

Jerger, Emmanuel Gene, b. 1925 in Eaton, Colorado; d. 2020 in Moorhead, Minnesota. Interview by Eric Mischnick in Moorhead, Minnesota, on 4 May 2008.

Johner, Jacob, b. 1922 in Scott, Saskatchewan; d. 2017 in Maidstone, Saskatchewan. Interview by Jessica Clark in Tramping Lake, Saskatchewan, on 14 July 2006.

Johner, Lawrence, b. 1920 in Kerrobert, Saskatchewan, d. 2017. Interview by Jessica Clark in Tramping Lake, Saskatchewan, on 14 July 2006.

Johner, Peter, b. 1914 in Tramping Lake, Saskatchewan, d. 2010 in Edmonton, Alberta. Interview by Jessica Clark in Tramping Lake, Saskatchewan, on 14 July 2006.

Johnson, Elsie (Just), b. 1919 in Cumberland, Wisconsin; d. 2019 in Fargo, North Dakota. Interview by Eric Mischnick and Cassie Ptacek in Fargo, North Dakota, on 28 April 2008.

Johnson, Lyla (Pressler), b. 1920 in Eureka, South Dakota; d. 2017 in Wishek, North Dakota. Interview by Sarah (Lacher) Crary in Lehr, North Dakota, on 13 August 2006.

Johnson, Ruth Rosa (Korus), b. 1921 in Metz Township, North Dakota; d. 2014 in Moorhead, Minnesota. Interview by Cassie Ptacek and Jessica Clark in Moorhead, Minnesota, on 5 May 2007.

Kaseman, Adam A., b. 1927 in McIntosh County, North Dakota; d. 2022 in Jamestown, North Dakota. Interview by Jessica Clark in Gackle, North Dakota, on 24 May 2005.

Kaseman, Lorraine (Meidinger), b. 1930 in Zeeland, North Dakota; d. 2016 in Jamestown, North Dakota. Interview by Jessica Clark in Gackle, North Dakota, on 24 May 2005.

Keen, Geraldine (Dukart), b. 1938 in Allan, Saskatchewan. Interview by Jessica Clark in Saskatoon, Saskatchewan, on 22 July 2006.

Ketterling, Eileen M. (Burlack), b. 1930 in Burnstad, North Dakota; d. 2021 in Wishek, North Dakota. Interview by Sarah (Lacher) Crary in Wishek, North Dakota, on 11 August 2006.

Kittler, Edwin, b. 1948 in Hazen, North Dakota. Interview by Cassie Ptacek in Beulah, North Dakota, on 17 June 2008.

Klein, Joe, b. 1927 in Zeeland, North Dakota; b. 2019 in Bismarck, North Dakota. Interview by Sarah (Lacher) Crary in Zeeland, North Dakota, on 7 August 2006.

Klein, Reinhold R., b. 1924 in Wishek, North Dakota; d. 2018 in Wishek, North Dakota. Interview by Jessica Clark in Wishek, North Dakota, on 17 July 2007.

Kleingartner, Lorraine (Mueller), b. 1921 in Dickey County, North Dakota; d. 2011 in Jamestown, North Dakota. Interview by Jessica Clark in Gackle, North Dakota, on 6 June 2005.

Kleingartner, Rupert, b. 1923 in Gackle, North Dakota; d. 2009 in Jamestown, North Dakota. Interview by Jessica Clark in Gackle, North Dakota, on 6 June 2005.

Klettke, Marie (Rall), b. 1922 in Redcliff, Alberta; d. 2012 in Wishek, North Dakota. Interview by Eric Mischnick in Kulm, North Dakota, on 30 May 2008.

Klundt, Leona (Miller), b. 1931 in Gackle, North Dakota; d. 2019 in Bismarck, North Dakota. Interview by Jessica Clark in Gackle, North Dakota, on 29 June 2005.

Knell, Victor, b. 1941 in Hazen, North Dakota. Interview by Andrea Mott in Fargo, North Dakota, on 1 May 2009.

Kocher, Peter, b. 1918 in Hague, North Dakota; d. 2020 in Strasburg, North Dakota. Interview by Sarah (Lacher) Crary in Hague, North Dakota, on 4 August 2006.

Koepplin, Ida (Zimmerman), b. 1915 in Lehr, North Dakota; died in 2018 in Mesa, Arizona. Interview by Jessica Clark in Lehr, North Dakota, on 11 July 2007.

Kohlman, Joseph George, b. 1938 in Cactus Lake, Saskatchewan; d. 2009 in Saskatoon, Saskatchewan. Interview by Jessica Clark in Tramping Lake, Saskatchewan, on 13 July 2006.

Konrad, Viola (Eslinger), b. 1922 in Fredonia, North Dakota, d. 2021 in Jamestown, North Dakota. Interview by Sarah (Lacher) Crary in Kulm, North Dakota, on 5 August 2006.

Kraft, George, b. 1909 in Strassburg, South Russia; d. 2008 in Rugby, North Dakota. Interview by Jessica Clark in Rugby, North Dakota, on 8 July 2006.

Kraft, Larry, b. 1937 in Strasburg, North Dakota. Interview by Cassie Ptacek in Fargo, North Dakota, on 12 May 2008.

Kraft, Peter A., b. 1917 in Strasburg, North Dakota; d. 2008 in Linton, North Dakota. Interview by Sarah (Lacher) Crary in Linton, North Dakota, on 13 May 2006.

Krismer, Christina Rose Marie (Bast), b. 1937. Interview by Jessica Clark in Regina, Saskatchewan, on 28 July 2006.

Kurtz, Balzer, b. 1928 in Devils Lake, North Dakota; d. 2008 in Grand Forks, North Dakota. Interview by Cassie Ptacek in Devils Lake, North Dakota, on 17 June 2007.

Lang, Elizabeth Barbara (Rissling), b. 1938 in Scott, Saskatchewan. Interview by Jessica Clark in Tramping Lake, Saskatchewan, on 15 and 16 July 2006.

Lang, Philip, b. 1934 in Tramping Lake, Saskatchewan; d. 2020. Interview by Jessica Clark in Tramping Lake, Saskatchewan, on 15 and 16 July 2006.

Lapp, Selma (Job), b. 1927 in McPherson County, South Dakota; d. 2023 in Eureka, South Dakota. Interview by Cassie Ptacek in Eureka, South Dakota, on 23 May 2008.

Laturnus, Rosa (Schneider), b. 1925 in Scott, Saskatchewan. Interview by Jessica Clark in Tramping Lake, Saskatchewan, on 11 July 2006.

Leier, Barbara (Mitzel), b. 1927 in Kintyre, North Dakota. Interview by Andrea Mott in Braddock, North Dakota, on 6 June 2009.

Leier, Joseph C., b. 1921 Kintyre, North Dakota; 2019 in Napoleon, North Dakota. Interview by Andrea Mott in Braddock, North Dakota, on 6 June 2009.

Leippi, Mona (Wetstein), b. 1934 in Regina, Saskatchewan. Interview by Jessica Clark in Kronau, Saskatchewan, on 26 July 2006.

Leippi, Wilfred W., b. 1933. Interview by Jessica Clark in Kronau, Saskatchewan, on 26 July 2006.

Long, Christina (Aberle), b. 1930 in Emmons County, North Dakota; d. 2015 in Bismarck, North Dakota. Interview by Jessica Clark in Berlin, North Dakota, on 15 July 2005.

Mathern, Joseph, b. 1924 in Strasburg, North Dakota; d. 2013 in Edgeley, North Dakota. Interview by Eric Mischnick in Edgeley, North Dakota, on 31 May 2008.

Meidinger, Clarence, b. 1929 in Lehr, North Dakota; d. 2016 in LaMoure, North Dakota. Interview by Jessica Clark in Edgeley, North Dakota, on 14 July 2007.

Meier, Joseph Michael, b. 1921 in Tramping Lake, Saskatchewan. Interview by Jessica Clark in Luseland, Saskatchewan, on 12 July 2006.

Merkel, William Adam Jr., b. 1925 in McPherson County, South Dakota; d. 2017 in Eureka, South Dakota. Interview by Eric Mischnick in Eureka, South Dakota, on 17 and 18 May 2008.

Miller, Dorothy (Grasser), b. 1926 in Regent, North Dakota; d. 2020 in Richardton, North Dakota. Interview by Andrea Mott in Richardton, North Dakota, on 21 June 2009.

Miller, Julius, b. 1918 in Taylor, North Dakota; d. 2016 in Dickinson, North Dakota. Interview by Cassie Ptacek in Dickinson, North Dakota, on 2 and 5 August 2007.

Miller, Magdalena, b. 1916 in Knox, North Dakota; d. 2013 in Rugby, North Dakota. Interview by Cassie Ptacek in Knox, North Dakota, on 21 June 2007.

Miller, Malvin W., b. 1938 in Zap, North Dakota; d. 2009 in Zap, North Dakota. Interview by Cassie Ptacek in Golden Valley, North Dakota, on 13 June 2008.

Miller, Rose (Boespflug), b. 1922 in Dickinson, North Dakota; d. 2014 in Dickinson, North Dakota. Interview by Cassie Ptacek in Dickinson, North Dakota on 2 and 5 August 2007.

Miltenberger, Elizabeth (Bosch), b. 1915 in Ness Township, North Dakota; d. 2010 in Rugby, North Dakota. Interview by Cassie Ptacek in Rugby, North Dakota, on 19 June 2007.

Mock, Carolyn "Carol" (Marquart), b. 1927 in Napoleon, North Dakota; d. 2013 in Bismarck, North Dakota. Interview by Andrea Mott in Napoleon, North Dakota, on 5 June 2009.

Mutschelknaus, Eva (Helfrich), b. 1916 in Glen Ullin, North Dakota; d. 2010 in Dickinson, North Dakota. Interview by Cassie Ptacek in Dickinson, North Dakota on 8 August 2007.

Nagel, Armelia "Millie" (Ibach), b. 1932 in Strasburg, North Dakota. Interview by Jessica Clark in Bismarck, North Dakota, on 18 July 2007.

Nagel, Peter, b. 1929 in Linton, North Dakota; d. 2024 in Bismarck, North Dakota. Interview by Jessica Clark in Bismarck, North Dakota, on 18 July 2008.

Nill, Roy F., b. 1926 in Dickey County, North Dakota; d. 2015 in Wishek, North Dakota. Interview by Jessica Clark in Kulm, North Dakota, on 12 July 2007.

Opp, Edmund, b. 1920 in McPherson County, South Dakota; d. 2020 in Aberdeen, South Dakota. Interview by Cassie Ptacek and Eric Mischnick in Eureka, South Dakota, on 22 May 2008.

Opp, Esther Christiana (Mertz), b. 1926 in Dawson, North Dakota; d. 2015 in Bismarck, North Dakota. Interview by Andrea Mott in Napoleon, North Dakota, on 9 June 2009.

Ost, Gilbert, b. 1930 in Beulah, North Dakota. Interview by Cassie Ptacek in Beulah, North Dakota, on 10 June 2008.

Presler, LeRoy H., b. 1929 in Gackle, North Dakota; d. 2020 in Fargo, North Dakota. Interview by Jessica Clark in Gackle, North Dakota, on 8 and 23 June 2005.

Presler, Vernon M., b. 1937 in Cashmere, Washington; d. 2019 in Kulm, North Dakota. Interview by Sarah (Lacher) Crary in Kulm, North Dakota, on 8 August 2008.

Rasch, Elda (Schultz), b. 1928 in Fredonia, North Dakota; d. 2010 in Bismarck, North Dakota. Interview by Jessica Clark in Bismarck, North Dakota on 25 July 2007.

Rath, Tillie (Deisz), b. 1931 in Emmons County, North Dakota; d. 2013 in Wishek, North Dakota. Interview by Cassie Ptacek in Wishek, North Dakota, on 1 June 2008.

Rath, Wilbert, b. 1924 in McIntosh County, North Dakota; d. 2014. Interview by Eric Mischnick in Wishek, North Dakota, on 1 June 2008.

Riehl, Ann (Gray), b. 1937 in Zelma, Saskatchewan. Interview by Jessica Clark in Allan, Saskatchewan, on 20 July 2006.

Riehl, John, b. 1937 in Zelma, Saskatchewan; d. 2016. Interview by Jessica Clark in Allan, Saskatchewan, on 20 July 2006.

Roesler, Emil, b. 1923 in Lehr, North Dakota; d. 2020 in Fargo, North Dakota. Interview by Jessica Clark in Lehr, North Dakota, on 13 July 2007; interview by Eric Mischnick in Lehr, North Dakota, on 4 June 2008.

Rost, Norma (Pahl), b. 1936 in Forbes, North Dakota. Interview by Sarah (Lacher) Crary in Kulm, North Dakota, on 6 August 2006.

Rost, Norman, b. 1930 in Edgeley, North Dakota; d. 2023 in Jamestown, North Dakota. Interview by Sarah (Lacher) Crary in Kulm, North Dakota, on 6 August 2006.

Rudolph, Orion Arlyn, b. 1938 in Fredonia, North Dakota. Interview by Eric Mischnick in Ashley, North Dakota, on 14 May 2008.

Schaffer, Emil E., b. 1917 in Logan County, North Dakota; d. 2010, Jamestown, North Dakota. Interview by Jessica Clark in Gackle, North Dakota on 23 June 2005 and 18 July 2008.

Schall, Nick Jr., b. 1924 in Rugby, North Dakota. Interview by Jessica Clark in Rugby, North Dakota on 2 July 2006.

Schauer, Irene (Bender), b. 1928 in Danzig, North Dakota; d. 2021 in Wishek, North Dakota. Interview by Jessica Clark in Wishek, North Dakota, on 14 July 2007.

Schell-Smith, Mary (Vetter), b. 1913 in Linton, North Dakota. Interview by Sarah (Lacher) Crary in Fargo, North Dakota, on 27 June 2006.

Schlecht, Esther (Kauk), b. 1918 in Lehr, North Dakota; d. 2013 in Napoleon, North Dakota. Interview by Jessica Clark in Streeter, North Dakota, on 27 June 2005.

Schlenker, Alvina (Nitschke), b. 1925 in Ashley, North Dakota; d. 2020 in Ashley, North Dakota. Interview by Cassie Ptacek in Ashley, North Dakota, on 19 May 2008.

Schlenker, Harold, b. 1924 in McIntosh County, North Dakota; d. 2014 in Ashley, North Dakota. Interview by Eric Mischnick in Ashley, North Dakota on 19 May 2008.

Schmidt, Arthur E., b. 1922 in Neudorf Township, North Dakota; d. 2014 in Wishek, North Dakota. Interview by Jessica Clark in Gackle, North Dakota on 10 and 16 July 2007.

Schneider, Joseph J., b. 1921 in Napoleon, North Dakota; d. 2020 in Dickinson, North Dakota. Interview by Cassie Ptacek in Dickinson, North Dakota, on 3 August 2007.

Schott, Alma (Janke), b. 1922 in Gackle, North Dakota; d. 2012 in Moorhead, Minnesota. Interview by Jessica Clark in Gackle, North Dakota on 25 May and 8 June 2005.

Schumacher, Eugenia (Vetter), b. 1927 in North Dakota; d. 2024 in Fargo, North Dakota. Interview by Sarah (Lacher) Crary in Linton, North Dakota, on 26 May 2006.

Schwab, Imogene (Schwahn), b. 1929 in Linton, North Dakota; d. 2016 in Strasburg, North Dakota. Interview by Sarah (Lacher) Crary in Strasburg, North Dakota, on 25 May 2006.

Schwan, Fran, b. 1937 in Rugby, North Dakota. Interview by Cassie Ptacek in Devils Lake, North Dakota, on 18 June 2007.

Schwan, Simon, b. 1924 in Rugby, North Dakota; d. 2023 in McVille, North Dakota. Interview by Cassie Ptacek in Michigan, North Dakota, on 14 June 2007.

Sehn, Valeria (Wagner), b. 1923 in Strasburg, North Dakota, d. 2023 in Mandan, North Dakota. Interview by Sarah (Lacher) Crary in Linton, North Dakota, on 19 May 2006.

Selensky, Edward Matthew, b. 1935 in Rugby, North Dakota. Interview by Cassie Ptacek in Rugby, North Dakota on 20 and 21 June 2007.

Selensky, Elaine Marie (Volk), b. 1940 in McHenry County, North Dakota. Interview by Cassie Ptacek in Rugby, North Dakota, on 20 June 2007.

Sellinger, Agatha "Peggy," b. 1918 in Dilke, Saskatchewan; d. 2016 in Regina, Saskatchewan. Interview by Jessica Clark in Regina, Saskatchewan, on 27 July 2006.

Sellinger, Mary, b. 1923 in Dilke, Saskatchewan; d. 2015 in Regina, Saskatchewan. Interview by Jessica Clark in Regina, Saskatchewan, on 27 July 2006.

Senger, Agnes (Brickner), b. 1914 in Colonsay, Saskatchewan; d. 2011. Interview by Jessica Clark in Allan, Saskatchewan, on 20 July 2006.

Senger, Mary Ann (Axtman), b. 1918 in Balta, North Dakota; d. 2011 in Devils Lake, North Dakota. Interview by Cassie Ptacek in Devils Lake, North Dakota, on 16 June 2007.

Sibley, Marian (Dick), b. 1917 in Englevale, North Dakota; d. 2015 in Fargo, North Dakota. Interview by Cassie Ptacek in Moorhead, Minnesota, on 12 June 2007.

Smith, Marlene Ann (Kraft), b. 1938 in Pierce County, North Dakota. Interview by Cassie Ptacek in Penn, North Dakota, on 15 June 2007.

Sperling, Elmer D., b. 1926 in Lehr, North Dakota; d. 2014 in Lehr, North Dakota. Interview by Jessica Clark in Lehr, North Dakota, on 13 July 2007.

Stern, Shirley Ann (Weigum), b. 1932 in Zap, North Dakota. Interview by Cassie Ptacek in Golden Valley, North Dakota, on 14 June 2008.

Thurn, Herbert, b. 1927 in Venturia, North Dakota. Interview by Jessica Clark in Bismarck, North Dakota, on 23 July 2007.

Thurn, Mildred D. (Dockter), b. 1931 in Venturia, North Dakota; d. 2017 in Bismarck, North Dakota. Interview by Jessica Clark in Bismarck, North Dakota, on 23 July 2007.

Tschaekofske, Cora (Wolff), b. 1925 in Golden Valley, North Dakota; d. 2012 in Bismarck, North Dakota. Interview by Cassie Ptacek in Dickinson, North Dakota, on 6 August 2007.

Urlacher, Eleanor (Hutmacher), b. 1932 in Dickinson, North Dakota. Interview by Andrea Mott in Dickinson, North Dakota, on 23 June 2009.

Urlacher, Emilia (Hutmacher), b. 1930 in Fayette, North Dakota. Interview by Andrea Mott in Dickinson, North Dakota, on 26 June 2009.

Urlacher, Frank, b. 1928 in Kennedy Township, North Dakota; d. 2016 in Dickinson, North Dakota. Interview by Andrea Mott in Dickinson, North Dakota, on 26 June 2009.

Urlacher, John D., b. 1932 in Kennedy Township, North Dakota; d. 2023 in Dickinson, North Dakota. Interview by Andrea Mott in Dickinson, North Dakota, on 23 June 2009.

Vetter, August, b. 1930 in Emmons County, North Dakota; d. 2017 in Bismarck, North Dakota. Interview by Sarah (Lacher) Crary in Linton, North Dakota, on 16 and 25 May 2006.

Vetter, James, b. 1938 in Logan County, North Dakota; d. 2021 in Fargo, North Dakota. Interview by Sarah (Lacher) Crary and Jessica Clark in Fargo, North Dakota, on 13 April 2006.

Vetter, Wendelyn, b. 1933 in Strasburg, North Dakota; d. 2016 in Grand Forks, North Dakota. Interview by Sarah (Lacher) Crary in Hillsboro, North Dakota, on 9 May 2006.

Wald, Catherine (Glatt), b. 1930 in Hague, North Dakota; d. 2018 in Strasburg, North Dakota . Interview by Sarah (Lacher) Crary in Strasburg, North Dakota, on 24 May 2006.

Wald, Felix, b. 1930 in Strasburg, North Dakota; d. 2018 in Strasburg, North Dakota. Interview by Sarah (Lacher) Crary in Strasburg, North Dakota, on 24 May 2006.

Wald, Mildred (Glatt), b. 1936 in Hague, North Dakota. Interview by Sarah (Lacher) Crary in Venturia, North Dakota, on 31 May 2006.

Wald, William "Bill," b. 1933 in Venturia, North Dakota; d. 2020 in Zeeland, North Dakota. Interview by Sarah (Lacher) Crary in Venturia, North Dakota, on 31 May 2006.

Wanner, Gust, b. 1916 in Wishek, North Dakota; d. 2010 in Wishek, North Dakota. Interview by Jessica Clark in Wishek, North Dakota, on 13 July 2005.

Weber, Mary (Mitzel), b. 1915 in Kintyre, North Dakota; d. 2008 in Bismarck, North Dakota. Interview by Sarah (Lacher) Crary in Hague, North Dakota, on 27 May 2006.

Welder, Paul P., b. 1913 in Strasburg, North Dakota; d. 2010 in Linton, North Dakota. Interview by Sarah (Lacher) Crary in Linton, North Dakota, on 15 May 2006.

Werth, Gorden R., b. 1931 in Lehr, North Dakota. Interview by Jessica Clark in Lehr, North Dakota, on 15 July 2007.

Wiederanders, Richard A., b. 1924 in Clinton, Iowa; d. 2018 in Eureka, South Dakota. Interview by Eric Mischnick in Eureka, South Dakota, on 18 May 2008.

Wolbaum, Harry "Ronamus" Jr., b. 1924 in St. Michael, North Dakota; d. 2019 in Mandan, North Dakota. Interview by Andrea Mott in Braddock, North Dakota, on 7 June 2009.

Wolbaum, Magdalena "Lena" (Jangula), b. 1929 in Braddock, North Dakota; d, 2020 in Bismarck, North Dakota. Interview by Andrea Mott in Braddock, North Dakota, on 7 June 2009.

Wolf, Rueben Richard, b. 1933 in Havelock, North Dakota; 2021 in Kulm, North Dakota. Interview by Cassie Ptacek in Kulm, North Dakota, on 5 June 2008.

Zeiler, Doreen (Brossart), b. 1932 in Allan, Saskatchewan. Interview by Jessica Clark in Allan, Saskatchewan, on 18 July 2006.

Zeiler, Reynold, b. 1931 in Allan, Saskatchewan; d. 2022 in Saskatoon, Saskatchewan. Interview by Jessica Clark in Allan, Saskatchewan, on 18 July 2006.

Ziegler, Dorothea Anna (Bergstedt), b. 1919 in Jordan, Montana; d. 2022 in Bismarck, North Dakota. Interview by Andrea Mott in Dickinson, North Dakota, on 29 June 2009.

Zimmerman, Delmar D., b. 1933 in Wishek, North Dakota; d. 2025 in Wishek, North Dakota. Interview by Jessica Clark in Wishek, North Dakota, on 19 July 2005; interview by Sarah (Lacher) Crary in Wishek and Lehr, North Dakota, on 12 August 2006.

Zimmerman, Deloris (Boschee), b. 1933 in Lehr, North Dakota; d. 2022. Interview by Jessica Clark in Wishek, North Dakota, on 19 July 2005; interview by Sarah (Lacher) Crary in Wishek and Lehr, North Dakota, on 12 August 2006.

Zuern, Elden D. J., b. 1928 in Hazen, North Dakota; d. 2020. Interview by Cassie Ptacek in Hazen, North Dakota, on 16 June 2008.

INTRODUCTION
The New German-Russian Story

"I was no little angel," says Donald T. Huber of Hazen, North Dakota, "but I was not a complete devil either!" On occasion, his pranks got him into trouble. Donald recounts the time he and some other German-Russian boys were playing cops and robbers after school and had fun at the expense of another boy, a social outcast. Playing on the desires of this unpopular boy, Donald and his friends convinced him to partake in their game. The children then tied this poor boy to a nearby tree and left him there. Around suppertime, the boy's mother knocked on Donald's parents' door, asking for the whereabouts of her son. Since the Huber farm was closest to the playground, it was the logical place to inquire first. Trying to save face, Donald lied, telling the boy's mother they had accidentally left her son tied to a tree. Sticking to his story years later, Donald exclaims, "I can't believe we forgot about him!"[6]

A lie in childhood, an exaggeration in memory, such as told by Donald, poses issues as to the veracity of oral histories (and other memory sources), as well as all historical documents, of course. Too often scholars deem these sources—oral histories and childhood narratives—as nothing more than stories or folk tales, and conclude they are not worthy of adding to the historical narrative. Yet, these stories reveal new perspectives on the past and the present. By sharing childhood memories, interviewees shed light on the lived experiences of overlooked participants from the past (i.e., children), as well as the construction of collective memory taking place in the present. This conversation between the past and the present makes the oral history methodology incredibly valuable to the studies of history and memory.

[6] Donald T. Huber, interview.

Throughout this work, I use terms such as history, memory, identity, ethnicity, childhood, and complexity. For the most part, I use definitions established by social and cultural historians who define these concepts as social constructions. As with any socially constructed idea, these terms are not historically static. They vary in place and time. Yet, the evolution of these ideas makes the study of the past fascinating and the need for term clarification essential. History, as defined in this work, is the interpretation of the past, using texts to discover and understand human experiences and ideas.[7] Historian Peter Burke describes memory as the "reconstruction of the past," while collective memory is the "memories constructed by social groups."[8] Identity, as interpreted by Burke, refers to self-perception, whether individual or collective.[9] Ethnicity is an imagined identity, according to historian Jon Gjerde, based on similarities in background, language, and religion.[10] Historian Hugh Cunningham argues that *childhood* is a constructed concept that differentiates children from adults, whereas the term *children* refers to individuals who inhabit this socially defined stage of life, shaped by their lived experiences.[11] Examining these ideas through an analytical narrative lens reveals a story of complexity, or in the words of Great Plains historian Thomas D. Isern, a multi-dimensional, organized, yet unnoticed, story—a revised story.[12]

Throughout my studies, I have researched the construction and reconstruction of childhood memories using the oral history methodology. From 2005 through 2010, I had an amazing opportunity to collaborate with Michael M. Miller, then-director of the Germans from Russia Heritage Collection, and with Isern, University Distinguished Professor and professor of history at North Dakota State University, to launch and coordinate a landmark oral history project, which we called the Dakota Memories Oral History Project

[7] For more information on the definition of history, see E.H. Carr, *What is History?* (Palgrave, 2001, 2nd ed.); Joyce Appleby, "The Power of History," *American Historical Review* (February 1998): 1–14.

[8] Peter Burke, "History as Social Memory," in *Varieties of Cultural History* (Ithaca: Cornell University Press, 1997), 44.

[9] Burke, *What Is Cultural History?* (Cambridge: Polity, 2008, 2nd ed.), 91.

[10] Jon Gjerde, *The Minds of the West: Ethnocultural Evolution in the Rural Middle West, 1830–1917* (Chapel Hill: The University of North Carolina Press, 1999), 59.

[11] Hugh Cunningham, *Children and Childhood in Western Society since 1500*, ed. John Morrill and David Cannadine, 2nd ed., *Studies in Modern History* (London: Pearson Longman, 2005), 2–3.

[12] Thomas D. Isern, "Agency, Complexity, Memory: A Scholarship for Western Places," *The Social Science Journal* 51 (2014): 3.

(henceforth referred to as Dakota Memories). Together, we designed this project—originally named the Germans from Russia Oral History Project—to document the childhood memories of the second- and third-generation Germans from Russia. Specifically, we wanted to preserve the memories of growing up German-Russian on the northern plains. Interested in documenting childhood in the early twentieth century, Dakota Memories oral historians interviewed Germans from Russia mostly between the ages of sixty and one hundred, which is to say German-Russians born before 1945. The memories collected for the Dakota Memories project are the heart of this work, which focuses on the reconstruction of childhood memories by Germans from Russia of the second and third generations.

Dakota Memories is truly a sophisticated oral history project, both in terms of techniques and technologies. Before launching this project, in 2005, I participated in a semester-long, weekly research seminar on the methodology of oral history taught by Kimberly Porter (former editor of *The Oral History Review*). In this seminar I became acquainted with foundational works of oral history: Paul Thompson's *The Voice of the Past*, David K. Dunaway and Willa K. Baum's *Oral History*, Michael Frisch's *A Shared Authority*, and Robert Perks and Alistair Thomson's *The Oral History Reader*. At the same time, Michael Miller (Dakota Memories Director) and I consulted with the director of Prairie Public television, Bob Dambach, for advice on documentary quality equipment, such as video cameras, tripods, microphones, and digital cameras.[13]

As per the requirements of our institution, North Dakota State University (NDSU), Miller, Isern, and I also needed to secure approval to work with human subjects. At the university, the Office of Research, Creative Activities, and Technology Transfer (RCATT) was responsible for overseeing all research projects involving human subjects. The first step in obtaining approval involved training sessions and certifications. According to university requirements, the project founders and interviewers needed certifications before making contact with human subjects. In addition, as per the RCATT requirements, the founders of Dakota Memories completed and submitted an application for an *exempt*

---

[13] Paul Thompson, *The Voice of the Past: Oral History* (Oxford: Oxford University Press, 2000); David K. Dunaway and Willa K. Baum, eds., *Oral History: An Interdisciplinary Anthology* (New York: AltaMira Press, 1996); Michael Frisch, *A Shared Authority: Essays on the Craft and Meaning of Oral and Public History* (New York: University Press of New York [SUNY], 1990); Robert Perks and Alistair Thomson, eds., *The Oral History Reader* (London: Routledge, 2004).

status protocol, Institutional Review Board for the Protection of Human Participants in Research. This application involved explaining the project's mission, scope, participants, methods, and techniques. Because this oral history project did not involve experimentation, minors, or pregnant women, the RCATT office approved the project's *exempt status*.

The mission of Dakota Memories was to preserve the history and heritage of the second- and third-generation Germans from Russia. We focused on childhood memories and family relationships. In short, we chronicled what it was like growing up German-Russian on the northern plains.[14]

The broad scope of the project is childhood, with several subtopics addressed during each oral history interview. We had questions pertaining to the environment, family relationships, pastimes and activities, grade school, chores and responsibilities, daily life, farm and community life, town and community life, religion, holidays and celebrations, youth activities, courtship practices, high school, health care, funerals, historical events, higher education, and weddings. Of course, narrators also shared stories dealing with their adult life. Nonetheless, the focus remains on the memories of growing up German-Russian on the northern plains. As Dakota Memories project founders, Miller, Isern, and I drafted specific questions for each of the subtopics (see Appendix A). Since the GHRC staff had been collecting interviews with this ethnic group for more than thirty years, we used their existing questions as a foundation. In many cases, the questions simply needed rewording, making them more open-ended and conducive to lengthy responses. We also organized the questions into the topical framework, such as the life history approach, modeled by Thompson in *The Voice of the Past*.[15]

Identifying the project's potential participants was easy: Germans from Russia on the northern plains. Finding interested participants, however, was a little more difficult, especially during the project's first year. Fortunately, Dakota Memories was associated with a long-standing heritage collection for the ethnic group—the GHRC—a reputable organization among German-Russian communities. Thus, the oral history project had credibility from the onset.

Despite this connection, approximately half the people invited to participate in the first season declined. People felt unqualified or ill equipped to par-

[14] Germans from Russia Heritage Collection, Dakota Memories Oral History Project, North Dakota State University Libraries, https://digitalhorizonsonline.org/digital/collection/p16921coll13.

[15] For more information, see Thompson, *The Voice of the Past*, 309–23.

ticipate, even though Miller and his associates had suggested them. They did not believe they had a story to tell. A few felt reluctant about having strangers in their homes or about being video recorded. Nonetheless, the Dakota Memories staff successfully completed season one in 2005. Seasons two through five were much easier. By the end of the first year, the project leaders had built a reputation among German-Russian communities, and people began volunteering themselves, as well as family members and friends, to participate. Thus, the memories of this self-selected group—the memory makers of German-Russians on the northern plains—are the basis of the Dakota Memories project.

Once potential narrators agreed to participate—which Dakota Memories oral historians usually arranged via cold calls, or phone conversations—the Dakota Memories staff sent out introductory packets. These packets included an introductory letter explaining the oral history process, the project and collection brochures, a biography form (see Appendix C), a topic checklist (see Appendix D), and a self-addressed, stamped envelope. Providing the self-addressed, stamped envelope helped ensure the return of all requested material, such as the biography form and topic checklist. The biography form closely resembled the family data form in Linda Shopes's essay, "Using Oral History for a Family History Project."[16] The checklist allowed narrators to familiarize themselves with topics that they might discuss during the oral history project. These forms provided the necessary information—such as topics, relevant dates, and proper spellings—the historians needed to best prepare for the individual interviews.

Oral historians of this project preferred to conduct their interviews on the narrators' home turf. In other words, we established scholarly residencies within each community. These residencies typically consisted of an oral historian and a videographer visiting a community for five to ten days, interviewing anywhere from five to twenty people depending on the time and locale. This arrangement involved teams immersing themselves in their interviewees' community life. The scholarly residency allowed interviewees, as well as the community, to become familiar with the oral historian and the videographer. The practice also permitted the oral history team to devote entire days to any given narrator. Not only was the project design cost effective by cutting down on travel expenditures, but it was also more beneficial to the project's ultimate goal:

---

[16] Linda Shopes, "Using Oral History for a Family History Project," in *Oral History: An Interdisciplinary Anthology*, edited by David K. Dunaway and Willa K. Baum (New York: AltaMira Press, 1996), 235.

the reconstruction of recollections. Nothing is more frustrating than leaving a narrator just when they are revealing their deep memories. Time and patience are keys to good interviews.

In addition to the scholarly residency, the oral history team usually conducted the interviews in the narrators' homes. This location typically made narrators feel at ease, thereby creating the necessary environment for a productive interview. In "Using an Oral History for a Family History Project," Shopes writes, "an interview is above all a social interaction."[17] If an interview is awkward and tense, she states, "no matter how carefully researched, it will be intellectually and personally unrewarding."[18] Thus, it is important to create a comfortable and engaging atmosphere. Dakota Memories oral history teams excelled at creating this environment, as they began most interviews in the living rooms or kitchens of their narrators' homes. The more comfortable and at ease narrators felt, the more information they revealed.

In any one oral history, the narration is the memory of an individual, as well as a social group. Sometimes the individual memories may conform to the collective memories, but not always. Thus, it is important to examine both if possible. An examination of the collective memory reveals the social ideals of a group, while an examination of the individual memory reveals the complex reality of the individual.[19]

Because the Dakota Memories staff were interested in learning about the individual as well as the collective memories, oral historians preferred to conduct interviews on an individual basis. It was ideal to have only the oral historian, the videographer, and the narrator present during the oral history process. When guests (i.e., siblings, children, or spouses) were present, the oral history teams experienced difficulties getting narrators to reveal their individual memories. Not only did guests constantly interrupt the interview process, adding their own stories or perspectives to the conversation, but also many times narrators sought the approval of their guests by asking them to verify or elaborate on their memories. Thus, the teams usually asked that no one else was physically present during the interview process.

---

[17] Shopes, "Using an Oral History," 234.

[18] Shopes, "Using an Oral History," 234.

[19] Jessica Clark, "Holocaust Education and Collective Memory in West Germany, 1945–1970" (master's thesis, North Dakota State University, 2004), 2–3; Burke, "History as Social Memory," 44.

Regardless, it is impossible to eliminate all external influences during an oral history. In *A Shared Authority*, Frisch acknowledges that participants in oral histories share the authority.[20] This means oral historians and narrators are constantly negotiating and renegotiating their authority throughout the course of an interview. For Dakota Memories oral historians, the sharing of authority was more complex: present during the interviews were the Dakota Memories team, a narrator, and the German-Russian community even though the latter was not physically in the room. Long before the oral history teams established their scholarly residencies in communities, oral history participants visited with family and friends, comparing experiences and—without intention—reinforcing their collective memory. This circumstance began as soon as people volunteered or agreed to participate. Thus, by the time oral historians sat down with their narrators, the German-Russian community had already influenced the stories of every participant.

Getting at the individual, cached memory required time, persistence, and occasionally stimuli. Time provided the opportunity for two things to unfold: establishment of a solid relationship and deeper inquiry. The formation of a solid relationship made narrators feel more comfortable and at ease; as a result, they tended to share deeper memories. Dakota Memories oral history sessions typically lasted many hours in order to tap into those memories, resulting in an average of three to four hours of video footage.[21]

Persistence was another key to getting past the collective memory via oral history. Alessandro Portelli, an oral historian, states, "oral sources tell us not just what people did, but what they wanted to do, what they believe they were doing, and what they now think they did."[22] It was the responsibility of Dakota Memories oral historians to stay committed to the task. At first narrators only shared superficial, one-dimensional stories. With time and with calculated persistence, however, these oral historians broke into the multi-dimensional stories—the deep memories. After narrators shared their initial memories, oral historians kept digging for the complexities by asking follow-up questions or by rephrasing previous questions. This process revealed more than the generally accepted collective story.

[20] For more information, see Frisch, *A Shared Authority*.

[21] For more information, see Hugo Slim, et al., "Ways of Listening," in *The Oral History Reader*, ed. Robert Perks and Alistair Thomson (London: Routledge, 2004).

[22] Alessandro Portelli, "What Makes Oral History Different," in *The Oral History Reader*, ed. Robert Perks and Alistair Thomson (London: Routledge, 2004), 67.

Using stimuli also helped the historians to get at the deep memories.[23] Thus, along with time and persistence they employed a number of special techniques (see Appendix B). Some of the special techniques employed by these oral historians to trigger or stimulate individual or deep memories were photo album tours, memory maps, landmark walks, cemetery walks, cooking demonstrations, and music demonstrations. These techniques not only helped the narrators to move past their collective narrative, but they also helped them forget about the intimidating video camera recording the interviews. A photo album tour—a special technique employed most often by the Dakota Memories team—involved an oral historian and a narrator using historic photographs to generate stories or spark memories. This technique was also a great tool for gaining archival documents to accompany the oral history narratives. The walks (landmark and cemetery) consisted of teams and narrators visiting places important to the story. Bringing narrators back to the environmental setting revealed familiar sights and sounds, and as a result, stories they may have otherwise overlooked. The cooking and music demonstrations involved recreating experiences in order to jog memories. The memory-mapping technique required narrators to draw a place from their past. These sensory memory triggers were effective. Combined with comfortable atmospheres and personable oral history teams, the triggers allowed Dakota Memories oral historians to gather fascinating memories.

Initially, narrators of the Dakota Memories only shared their collective story with their oral historians. The resulting narrative mirrors the story memoirist Pauline Neher Diede recounts in *Homesteading on the Knife River Prairies*. In this fascinating memoir, Diede shares stories of her family's immigration to, and life in, America. She, in line with the time-honored German-Russian story, tells a tale of persecution, hardships, isolation, and sadness. In fact, throughout this eighty-eight-page memoir, Diede hardly mentions any enjoyable or positive experiences. *Homesteading on the Knife River Prairies* is thus a true representation of the German-Russian first-generation's traditional narrative.[24]

---

[23] Slim, et al., "Ways of Listening," 120–22.

[24] Pauline Neher Diede, *Homesteading on the Knife River Prairies*, ed. Elizabeth Hampsten (Bismarck, ND: Germans from Russia Heritage Society, 1983), 7, 17–18, 20, 22, and 30.

The immigrant generation of Germans from Russia did not leave their homeland by choice—at least, not in the way many European immigrants sought opportunity in America. They left because they felt forced, their place in Russian society becoming increasingly untenable by the late nineteenth century. Their ancestors had originally migrated to Russia at the invitation of Catherine the Great, lured by promises of religious freedom, land, and autonomy. For decades, they thrived in tight-knit, German-speaking villages. But as Russian nationalism intensified, those same privileges were stripped away. Military conscription, land shortages, and restrictions on language and education made life increasingly difficult. Faced with dwindling options, many Germans from Russia saw no future for themselves in the empire and reluctantly made the difficult decision to uproot once again—leaving behind not only their homes but generations of cultural and communal identity.[25]

Their new homeland offered little comfort. Unlike immigrants who arrived in America seeking fortune in cities or industrial centers, Germans from Russia found themselves scattered across the Great Plains, where the land was as unrelenting as their outsider status. The prairie demanded everything of them—backbreaking labor, perseverance, and an acceptance of isolation that few had anticipated. Many German-Russians settled in near-total social exclusion, surrounded by other German-speaking families but distanced from the broader American society that viewed them with suspicion. Their unfamiliar dialects, distinct customs, and steadfast commitment to preserving their culture made them targets of prejudice, particularly during World War I, when anti-German sentiment swept through the United States. Even within immigrant communities, Germans from Russia found themselves separate—not quite Russian, not quite German, and not quite American. Hard work and endurance became their defining characteristics, but even as they forged new lives, many held on to a sense of dislocation. Success, for this first generation, was not about prosperity but survival—making a home in a land that had not yet welcomed them.[26]

---

[25] Richard Sallet, *The Germans from Russia in the United States* (Minneapolis: University of Minnesota Press, 1974), 3–12; Renate Bridenthal, *The Heimat Abroad: The Boundaries of Germanness* (Ann Arbor: University of Michigan Press, 2008), 187–218; Timothy J. Kloberdanz, "The Volga Germans in North Dakota," *North Dakota History* 65, no. 2 (1998): 30–45.

[26] Diede, *Homesteading on the Knife River Prairies*, 7, 17–18, 30; David L. Danbom, *Born in the Country: A History of Rural America* (Baltimore: Johns Hopkins University Press, 2006), 102–15; Dakota Memories Oral History Project, NDSU Libraries, *Digital Horizons*, https://digitalhorizonsonline.org/digital/collection/p16921coll13.

Yet, the narrations of the Dakota Memories participants reveal that there is more to the story of German-Russian descendants on the northern plains. While the second- and third-generation Germans from Russia did indeed inherit their ancestors' narrative, their personal experiences in their *heimat* (homeland) significantly altered the German-Russian story. During the Dakota Memories interviews, narrators reminisced about treasured memories of fun, adventure, and opportunities. For the second and third generations the story is no longer just about the persecutions and hardships; their story is now one of opportunities and success. Life on the northern plains, as German-Russian descendants insist, was not all that bad. They had plenty of positive experiences to balance with the negative experiences.

Getting the second- and third-generation German-Russians to share memories that may contradict their ethnic group's collective story takes finesse. In an article published by the *Oral History Review*, Lenore Layman acknowledges that it is difficult when narrators are reluctant to share stories, especially when the story veers away from the narrators' collective memory. It is an art, she says, to get narrators to move past the socially accepted shared memory. It requires oral historians and narrators to negotiate and renegotiate their relationship constantly throughout the oral history. Layman notes that reticence occurs when a "disjuncture [exists] between the individual memory and public or collective memory."[27] She continues, "[W]here individual memory differs from these patterns and exemplars, narrators are likely to fall reticent, perceiving the likelihood of disbelief or disapproval, or simply to avoid appearing disagreeable or offensive."[28] For narrators of the Dakota Memories, reticence exists because they fear losing or altering their ethnic heritage or narrative. These narrators participate in the Dakota Memories for the sake of ethnic group affirmations, and for them that means remembering and honoring the trials and tribulations of their ancestors.

Working together during the course of lengthy interviews, oral historians and narrators of Dakota Memories have pushed past the accepted collective story of Germans from Russia on the northern plains. Layman brings up the question regarding the ethics of oral historians "push[ing]" a narrator to share memories.[29] She questions whether "the end (in this case enlarging the histori-

---

[27] Lenore Layman, "Reticence in Oral History Interviews," *Oral History Review* 36, no. 2 (2009): 210–21.

[28] Layman, 21.

[29] Layman, 225–26.

cal record) justifies the means (forcing an unwilling person to comply)."[30] While Dakota Memories oral historians do indeed try to move past the collective story, they never force narrators to discuss anything. Furthermore, at any point the narrators may ask for a break, end the interview, or request something be removed or locked from the recording. Dakota Memories oral historians simply guide their narrators past the shared stories—the stories narrators think are historically relevant—and toward the individual stories, the stories narrators think are historically irrelevant.

Documenting and analyzing memories outside the established narrative proves difficult, as people generally dislike going against tradition. In an article originally published in the 1994 edition of the *Oral History Association of Australian Journal,* Naomi Rosh White examines silences in the narrations of Holocaust survivors. While the experience of a Holocaust survivor and a German-Russian descendant were vastly different, the issues of silence because of shame, guilt, or embarrassment have many similarities. She insists oral historians need to be patient, and occasionally employ memory triggers, in order to move narrators past traditions, as well as silence or avoidance. Then again, as White also acknowledges, sometimes the story is in the silence itself.[31]

Silence, as English professor Lawrence L. Langer explores in his 1991 work, *Holocaust Testimonies,* occurs for a number of reasons. Langer, in fact, discusses three forms of Holocaust testimonies that are applicable to a discussion of the memories of German-Russian descendants: "deep memory," "humiliated memory," and "unheroic memory."[32] Deep memories are individual memories that may differ whether slightly or significantly from the collective memory. Humiliated memories are memories of humiliating or embarrassing events. Unheroic memories are memories of selfishness and shame.[33] These types of memories create silence in testimonies as well as oral histories.

Oral historians of the Dakota Memories had to work around the barriers of collective memories and individual memories, but they also had to work around the problems of childhood memories. For instance, reconstructing infancy (birth to two years of age) is beyond difficult for narrators, as well as most

---

[30] Layman, 226.

[31] Naomi Rosh White, "Marking Absences: Holocaust Testimony and History," in *The Oral History Reader*, ed. Robert Perks and Alistair Thomson (New York: Routledge, 2004), 173–74.

[32] Lawrence L. Langer, *Holocaust Testimonies: The Ruins of Memory* (New Haven: Yale University Press, 1991), 1–38, 77–120, and 62–206.

[33] Langer, *Holocaust Testimonies.*

people, in general. Yet, as psychologist Daniel L. Schacter states in *Searching for Memory*, "young infants can retain specific details of particular episodes."[34] The problem lies in the fact that "prelinguistic infants" cannot share memories.[35] He says that before eight to nine months of age infant memories are "implicit" rather than "explicit."[36] Thus, people retain few, if any, memories prior to the age of two.

Memories from early childhood (three to five years of age) are less problematic yet are still vague. Schacter acknowledges that "older children remember much more than the younger ones."[37] Oral historian Paul Thompson even acknowledges that childhood memories tend to be "blurred and patchy and a bit confused."[38] The problematic nature of early childhood memories results from the memory capabilities of young children, as well as the temporal distance from the lived experience. Schacter points out that in this stage, children are just beginning to develop their linguistic skills. Furthermore, he maintains, with the development of language skills, children are learning "to impose narrative structure on their experiences."[39] Thus, memories from early childhood are spotty in part because at this stage children are just learning to construct their narratives.[40]

Memories of childhood (six to twelve years of age) are more vivid and detailed than memories of infancy and early childhood. Children at this stage have the capability to remember, as well as the ability to construct a narrative. Thus, there is more material to investigate. Yet, memory is still tricky. In *Childhood on the Farm*, historian Pamela Riney-Kehrberg states, "reflecting on childhood from some great distance may lead individuals to sugarcoat or distort some memories and fabricate others or omit important facts."[41] Yet, as Elliott West, a childhood and environmental historian of the American West, states,

---

[34] Daniel L. Schacter, *Searching for Memory: The Brain, the Mind, and the Past* (New York: Basic Books, 1996), 173.

[35] Schacter, *Searching for Memory*, 174.

[36] Schacter, *Searching for Memory*, 174.

[37] Schacter, *Searching for Memory*, 174.

[38] Paul Thompson, "Believe it or Not: Rethinking the Historical Interpretation of Memory," in *Memory and History: Essays on Recalling and Interpreting Experience*, ed. Jaclyn Jeffery and Glenace Edwall (London: University Press of America, Inc., 1994), 2.

[39] Schacter, *Searching for Memory*, 175.

[40] Schacter, *Searching for Memory*, 175.

[41] Pamela Riney-Kehrberg, *Childhood on the Farm: Work, Play, and the Coming of Age* (Lawrence: University Press of Kansas, 2005), 8.

"the only thing more foolish than accepting all this at face value would be to ignore it."[42] Childhood memories reveal much about lived experiences, as well as perceptions of the past.[43]

The closer someone is to the memory encoding, the more detail he or she will retain. Thus, memories of adolescence (thirteen years old to the time of marriage) tend to be more detailed than memories of infancy, early childhood, and childhood years.[44] Some narrators of the Dakota Memories, however, shy away from discussions of this transitional stage. Given that adolescence is, as childhood historian Steven Mintz writes, "a phase of emotional upheaval and fluctuating emotions marked by contradictory tendencies toward hyperactivity and inertness, selfishness and altruism, bravado, and a sense of worthlessness," many narrators have memories that Langer would categorize as deep, humiliated, or unheroic.[45] In other words, memories from this stage of life may reveal behaviors that were socially unacceptable or traumatic, and as a result, some people may be too embarrassed or reluctant to discuss their adolescent years. A few narrators of the *Dakota Memories*, for instance, even requested that their oral historian avoid specific topics or that video editors remove certain stories from their oral histories. Regardless, memories about adolescence (as well as the silence or avoidance of certain subjects) reveal the complexities of the German-Russian story.

At the heart of this work are these individual and collective childhood memories. The oral histories archived in the Dakota Memories office provide a wealth of insight into these memories, and as a result, historical topics that historians (and various scholars) have long ignored. Using oral histories, as Alessandro Portelli writes in *The Order Has Been Carried Out*, sheds light on the "personal and private feelings" of people.[46] It gives voice to the underrepresented. Furthermore, oral histories are living histories. Using this methodology, historians can interrogate the past directly and discover a wealth of secrets. "Each time an old person dies," says Portelli, "a library is burned."[47] Thus, oral

---

[42] Elliott West, *Growing Up with the Country: Childhood on the Far Western Frontier* (Albuquerque: University of New Mexico Press, 1989), xx.

[43] Riney-Kehrberg, *Childhood on the Farm*, 9.

[44] Schacter, *Searching for Memory*, 175.

[45] Steven Mintz, *Huck's Raft: A History of American Childhood* (Cambridge: Belknap Press of Harvard University, 2004), 196; Langer, *Holocaust Testimonies*, 1–38, 77–120, and 162–206.

[46] Alessandro Portelli, *The Order Has Been Carried Out: History, Memory, and Meaning of a Nazi Massacre in Rome* (New York: Palgrave, 2003), 16.

[47] Portelli, *The Order Has Been Carried Out*, 19.

history projects, such as the Dakota Memories, are saving one archival source (or narrative) at a time.

This study reveals a great deal about the studies of oral history, collective and individual memories, and childhood history. With regard to oral history, the study reveals that oral historians can tap into new stories and memories by employing the best, established oral history practices. Regarding collective and individual memories, this study reveals that deep-interviewing techniques license narrators to change their established collective memory by encouraging them to share deep, individual memories. As for childhood history, this study reveals that an examination of ethnicity sheds new light on the history of childhood in the American West.

Employing the model established by previous childhood and oral historians, such as Elliott West, Pamela Riney-Kehrberg, and Marilyn Irvin Holt, this work takes a topical and narrative approach to the story of growing up German-Russian on the northern plains.[48] For Germans from Russia of the second and third generations, childhood spans from infancy to marriage; thus, this book addresses memories from birth to the early twenties. There are chapters devoted to place, family relationships, chores and responsibilities, grade school, pastimes and activities, religion, holidays and celebrations, and coming of age. Weaved within each chapter are the voices of the narrators. Using narratives for analysis provides a clear window into the past, as well as the present.

Chapter one, "Reconstructing a Story," traces the evolution of the fields of oral history, memory studies, ethnic studies, and childhood history. This chapter concludes with an overview of rural childhood from the Progressive Era through post-World War II (1870–1960), establishing the background of German-Russian childhood in America and Canada. In essence, this chapter places the book firmly within these established discussions and contexts, revealing that this research not only reaffirms some of the existing literature, but also adds original findings.

Chapter two, "Amber Waves," examines the role of children's interactions with the northern plains in shaping the memories of German-Russian descendants. This place gave these children opportunities for fun and enjoyment, and it contributed to their most trying and terrifying times. This chapter argues

[48] For more information, see West, *Growing Up with the Country*; Riney-Kehrberg, *Childhood on the Farm*; Marilyn Irvin Holt, *Children of the Western Plains: The Nineteenth-Century Experience* (Chicago: Ivan R. Dee, 2003).

that memories of place allow the second and third generations to confirm a traditional narrative—one of hardship and oppression—and to generate a new narrative, one of good times and opportunities.

Chapter three, "More than a Photo," focuses on the family unit. Narrators reminisce about the family relationships of grandparent-grandchild, parent-child, and sibling-sibling. Through these relationships, German-Russian descendants learned about their ethnic group's traditional narrative and reshaped it into a new story, blending their past with their present. Within the context of family relationships, these descendants forged a new German-Russian story rooted in their ethnic heritage and history, but serviceable to new, American- or Canadian-born generations.

Chapter four, "Work Makes Life Sweet," points out that like their parents and grandparents, the second- and third-generation German-Russians insist their life was hard. According to their collective memory, children worked hard, yet as their individual memories reveal, their hard work provided them with vast opportunities and memorable life experiences. Interweaving their bitter memories with their sweet memories of chores and responsibilities was essential for the maintenance of their ethnic identity and collective memory.

Chapter five, "Reading, Writing, and Arithmetic," reveals the vividness of grade-school memories. Schools, as narrators of the Dakota Memories project attest, were agents of acculturation. These narrators insist schools promoted becoming American or Canadian. The reality, however, was more complex. This chapter concludes that the one-room country schools actually reinforced ethnic identity, as they brought immigrant children together in homogenous groups, unintentionally contributing to German-Russians maintaining their ethnic identity and collective memory.

Chapter six, "Captain, May I?" looks at the role of pastimes and activities in German-Russian childhood. Dakota Memories narrators initially sidestepped stories of fun and enjoyment as their personal agendas focused on reestablishing the ethnic group's collective memory. After time and deeper inquiry, however, the narrators revealed stories about their childhood fun, stories about their enjoyable, yet educational activities. These are the memories that allow them to treasure their childhoods.

Chapter seven, "*Gott ist die Liebe*," examines the role of religion in German-Russian childhoods. Regardless of their religious denominations, Protestant or Catholic, churches aided in the resilience of German-Russian ethnicity.

This social institution allowed Germans from Russia to blend their ethnic identity by incorporating new elements into their existing identity. In other words, they were redefining their ethnicity through their religious activities.

Chapter eight, "Hope for Tomorrow," provides a glimpse into the holidays and celebrations of Germans from Russia. The holidays, as one narrator so eloquently states, provided German-Russian children with a "hope for tomorrow."[49] In the Dakota Memories archives there are detailed accounts of the holiday traditions observed by the immigrant generation and its descendants. Stories of these traditions reveal the descendants' desire for change, as the new narrative has more optimism and less fatalism than the original narrative.

Chapter nine, "A Fun Time," describes the coming-of-age process for German-Russian descendants. Their teenage years were not much different from others growing up in the rural areas of the northern plains. Their coming of age was a liberating experience, if not the most pleasurable for teenagers. This chapter reveals that the German-Russian story is no longer solely a tale of hardships, isolation, and misery; the new narrative includes stories about double dating, playing organized sports, and celebrating love.

Chapter ten, "We Didn't Know Any Better," wraps up the book by reiterating its discoveries in, or contributions to, the fields of oral history, memory studies, and childhood history. It reveals that there is significant complexity involved within the fields of memory studies and childhood history, and that the best way to get at these complexities is to employ the established practices and techniques of deep interviewing. Using oral history methodology allows historians, like the Dakota Memories oral historians, to interrogate the past, bringing forth stories of change and resilience.

The nearly two hundred German-Russians who participated in the Dakota Memories project demonstrated their desire to serve as their ethnic group's memory makers. Most of these narrators begin their tale with the traditional German-Russian story, stories of hardship and oppression. Yet narrators and trained oral historians of the Dakota Memories project pushed past this established narrative and into the deep, individual memories. These oral histories reveal that the new German-Russian story is rooted in an ancestral past with tales of trials and tribulations, shaped by childhood experiences and memories as well as present conditions. As told by narrators, this new German-Russian story is complex and reflective of generational change and ethnic resilience.

---

[49] Interview with Emmanuel Bertsch.

CHAPTER 1

Reconstructing a Story

Mildred (Glatt) Wald's favorite memory growing up was playing in the family's old Chrysler car body. "This was our playhouse," she declares. Mildred remembers in the springtime her older brother, Pete, always brought home the baby rabbits he caught while plowing in the fields. She loved keeping them in the Chrysler. Thinking back, Mildred treasures her memories of these rabbits. She recalls spending a good deal of time scavenging the countryside searching for grass to feed her new pets and using old sardine cans as water or milk dishes. Although the baby rabbits always died, Mildred concludes, "It was fun while it lasted."[1]

Mildred is one of nearly two hundred German-Russian descendants from the northern plains who participated in a substantial oral history project—the Dakota Memories Oral History Project (Dakota Memories). Like most narrators, Mildred retraces her childhood by reminiscing about a variety of topics including the environment, family relationships, chores, and pastimes. Simultaneously, she takes care to reaffirm her ethnic identity. Mildred's oral history is representative of most of the oral histories in this collection. Her version reveals her maintaining the collective memory of Germans from Russia, as well as her individual childhood memories of growing up on the northern plains. By participating in Dakota Memories, German-Russians of the second and third generations vigorously assert their ethnic identity as established by collective memory. Persistent and perceptive interviewing, however, brings forth rich recollections of complexity, reflective of generational change and ethnic resilience.

[1] Interview with Mildred (Glatt) Wald.

To understand memory and identity, historians rely on sources created from personal recollections, such as oral histories and memoirs. Oral history, as understood by scholars such as Linda Shopes, is a research-based approach to documenting these recollections through recorded interviews, structured conversations, or written narratives. Unlike oral tradition, which involves the transmission of collective stories, customs, and beliefs across generations, oral history captures firsthand experiences as individuals remember them. It is both a method and a source of historical inquiry, emphasizing the ways people interpret and convey their past. By preserving these voices, oral history provides valuable insights into personal narratives and broader historical contexts that might otherwise be lost. [2]

The field of oral history emerged in the late 1930s and early 1940s with the Federal Writers' Project (FWP) of the Works Progress Administration (WPA). The Historical Data Project Files, such as the ones used by childhood historian Elizabeth Hampsten in *Settlers' Children*, were part of the FWP.[3] However, because the FWP primarily aimed to provide employment rather than produce scholarly research, it did not employ professional historians or journalists. The lack of trained interviewers, combined with the absence of recording devices, raises questions—such as those posed by Shopes—about the reliability and accuracy of these early oral histories.[4]

These interviews of the 1930s, however, predate the professionalization of the oral history field. Historians have identified Allan Nevins as the first oral historian. Shope notes that "Nevins was the first to initiate a systematic and disciplined effort to record on tape, preserve, and make available for future research recollections deemed of historical significance."[5] She believes that Nevins's first interview in 1948 set off a chain of events, starting with the establishment of the Columbia Oral History Research Office and the creation of the oral history field. In the early years of this field—the 1950s—oral historians were concerned with interviewing only the elites. In the 1960s and 1970s, with the rise of new social movements such as the women's liberation movement, the counterculture movement, and the civil rights movement, historians began

---

[2] Linda Shopes, "What Is Oral History? History Matters: The US Survey on the Web" http://historymatters.gmu.edu.

[3] For more information, see Elizabeth Hampsten, *Settlers' Children: Growing up on the Great Plains* (Norman: University of Oklahoma Press, 1991).

[4] Shopes, "What Is Oral History?"

[5] Shopes, "What Is Oral History?"

incorporating the other classes in their narratives. This transition marks the official development of the oral history field, as historians began to "record the voices of the historiographically—if not historically—silent."[6]

In *The Voice of the Past*, historian Paul Thompson acknowledges the importance of this grassroots approach. He argues that one of (if not the) most significant contributions oral history has made to the field of history is giving a voice to ordinary people.[7] This innovative field contributed to the rise of ethnic studies and childhood history in America. It also revolutionized the field of rural history. By employing the methodology of oral history, historians can now examine a much-overlooked historical group: the children of immigrants in rural America. While historians have explored this group through documentary sources, oral history allows for interactive engagement, enabling researchers to ask follow-up questions, clarify meaning, and capture the complexities of personal reflection in ways that static documents cannot.[8]

Beyond its role in recovering the voices of overlooked individuals, oral history provides a window into how memory and identity are shaped over time. Rather than focusing solely on lived experiences, oral history captures the ways in which people reconstruct the past, articulate their identities, and interpret their own histories. These sources reveal both individual and collective expressions of identity, shedding light on how people shape their own historical narratives. More than just a method of preservation, oral history serves as a form of historical agency, allowing individuals—regardless of age, race, ethnicity, gender, or social status—to participate actively in the historical record by telling their own stories. Memory sources clearly reveal a connection between the past and the present. Given that history is the discovery and interpretation of the past, using texts to understand human experiences, oral historians must understand this relationship between the past and the present.[9] For instance, in 1991, Hampsten published an overly depressing work on childhood in North Dakota (*Settlers' Children*) relying primarily upon the WPA's Historical Data Project

---

[6] Shopes, "What Is History?"

[7] Thompson, *The Voice of the Past*, 3.

[8] For a study of immigrant children using documentary sources, see Molly P. Rozum, Grasslands Grown: Creating Place on the U.S. Northern Plains and Canadian Prairies (Lincoln: University of Nebraska Press, 2021).

[9] For more information on the definition of history, see E.H. Carr's *What Is History?* 2nd ed. (New York: Palgrave, 2001).

Image 1.1. Reviewing family photographs helped to trigger memories among the Dakota Oral History Project narrators like Calvin K. Fercho, who says of this photograph: "Here is a very handsome, good looking little fellow, who looks like he might be about two-years-old sitting on a piano bench," exemplifying how early snapshots connected descendants to their childhood identities. *Photograph, circa 1927, used with permission from the Calvin K. Fercho Collection, located in the Dakota Memories Oral History Project, Germans from Russia Heritage Collection, North Dakota State University Libraries, Fargo, North Dakota, hereafter referred to as GRHC.*

Files.[10] This project was part of President Franklin Delano Roosevelt's New Deal programs, designed to help in the recovery, relief, and reform of America during the Great Depression. Recent memories of the Depression and hard times colored the WPA interviewees' memories, and, as a result, contributed to an overly negative portrayal of childhood in the West. Thus, historians must take into account this dialogue between the past and the present when using memory sources to write their historical narratives.

Like all historical sources, memory sources have their weaknesses. Historian Elliott West says, "people never recall everything exactly as it was and often their accounts are wildly inaccurate. Memories of the distant past, especially youthful years, can be the most distorted of all."[11] Historian Lori Lah-

[10] For more information, see Elizabeth Hampsten, *Settlers' Children: Growing Up on the Great Plains* (Norman: University of Oklahoma Press, 1991).

[11] Elliott West, *Growing Up with the Country: Childhood on the Far Western Frontier* (Albuquerque: University of New Mexico Press, 1989), xx.

lum arrives at similar conclusions about memory sources, yet acknowledges, "dismissing them because adults wrote or spoke about their childhood merely serves to silence the voices of many."[12] In other words, as West states, "the only thing more foolish than accepting all this at face value would be to ignore it."[13]

Because this book focuses on the reconstruction of the German-Russian story, memory sources (in particular oral histories) are the ideal sources. Taking David M. Wrobel's advice "to move beyond any inclination to merely examine the accuracy of people's memories for the purpose of assessing the validity or truthfulness of their accounts," this work is a study of memory.[14] It is not a study of the lived experiences of Germans from Russia, rather, it focuses on the reconstruction of collective and individual memories through oral histories. It is the new German-Russian story as told by this ethnic group's self-selected memory-makers.

The concept of ethnicity, specifically the German-Russian identity, is rooted in collective memory, as it is an imagined construct based on the collective memory of an immigrant group. "Imagined common immigrant pasts," writes historian Jon Gjerde, "based on similar linguistic, national, or religious heritages, provided the foundation for the 'creation' of ethnic groups in the United States. In a process labeled 'ethnicization' by scholars, immigrant groups became ethnics."[15] Thus, a group's ethnic identity is rooted in its collective memory, or the construction of memories by social groups. As French sociologist Maurice Halbwachs acknowledges, individuals remember, but social groups determine what is memorable, and "how it will be remembered."[16] Memories are therefore both public and communal, because people reconstruct them by relying on what others have said and done. Individual memories amalgamate in the formation of a society's collective memory, which reduces the diversity

---

[12] Lori Ann Lahlum, "Growing Up in Norwegian-American Communities: A Preliminary Study of Childhood, Adolescence, and Young Adulthood," in *Norwegian-American Essays 2008:"Migration and Memory,"* ed. Øyvind Tveitereid Gulliksen and Harry T. Cleven (Oslo, Norway: NAHA, 2008), 109.

[13] West, *Growing Up with the Country*, xx.

[14] David M. Wrobel, *Promised Lands: Promotion, Memory, and the Creation of the American West* (Lawrence: University Press of Kansas, 2002), 12.

[15] Gjerde, *The Minds of the West*, 242.

[16] Burke, "History as Social Memory," 44; Maurice Halbwachs, *On Collective Memory*, ed. Lewis A. Coser, trans. Lewis A. Coser (Chicago: University of Chicago Press, 1992), 12; Halbwachs, *The Collective Memory*, trans. Francis J. Ditter Jr. and Vida Yazdi Ditter (New York: Harper Colophon Books, 1950), 12, 24, and 51; Clark, "Holocaust Education and Collective Memory in West Germany, 1945-1970," 2–3.

of recollections.[17] This shared memory allows immigrant groups to imagine a common past and, as a result, an ethnic identity.[18] Yet, as oral historian Alessandro Portelli writes, "the tale and the memory may include material shared with others, but the rememberer and the teller are always individual persons who take on the task of remembering and the responsibility of telling."[19]

For more than fifty years, scholars have examined the construction and evolution of ethnic identity in immigrant groups throughout America and Canada. The early works focused primarily on an exploration of assimilation. For instance, scholars Oscar Handlin and Frederick C. Luebke portray immigrants as passive victims to outside influences.[20] With the emergence of the history from below approach, in the 1980s and 1990s, historians began to revisit this established narrative. Works by Kathleen Neils Conzen, Jane Marie Pederson, and Gjerde, for instance, focus on the persistence and maintenance of ethnic identity. These scholars conclude that external forces, such as space and time, allowed ethnic groups to maintain their identities.[21]

This dichotomy of ethnic assimilation versus persistence, however, is not applicable to every ethnic group. Nuances exist, for instance, in analyzing German-Russian descendants through this dichotomy. They neither abandoned their ethnic identity, nor resisted adapting to their new nationalities. Rather, the German-Russian experience, as the memory-makers demonstrate, reflects a process of cultural diffusion. According to Timothy J. Kloberdanz, cultural diffusion is the act of freely borrowing cultural traditions rather than assimilating into mainstream society.[22] In other words, Germans from Russia chose to blend elements of their new environment into their existing identity. This process of

---

[17] Clark, "Holocaust Education and Collective Memory in West Germany, 1945–1970," 3.

[18] For more information, see Benedict Anderson, *Imagined Communities: Reflections on the Origin and Spread of Nationalism,* revised ed. (London: Verso, 1991).

[19] Portelli, "What Makes Oral History Different," 14.

[20] For more information, see Oscar Handlin, *The Uprooted,* 2nd ed. (Boston: Little, Brown and Company, 1973); Frederick C. Luebke, "Ethnic Group Settlements on the Great Plains," *Western Historical Quarterly* 8 (1977): 405–30; Luebke, ed. *Ethnicity on the Great Plains* (Lincoln: University of Nebraska, 1980).

[21] For more information, see Kathleen Neils Conzen, "Historical Approaches to the Study of Rural Ethnic Communities," in *Ethnicity on the Great Plains*, ed. Frederick C. Luebke (Lincoln: University of Nebraska Press, 1980), 1–18; Jane Marie Pederson, *Between Memory and Reality: Family and Community in Rural Wisconsin, 1870–1970* (Madison, WI: University of Wisconsin Press, 1992); Gjerde, *The Minds of The West.*

[22] Timothy J. Kloberdanz, "'The Traditions We Share': Folklore in Four Cultures on the Northern Plains," *North Dakota History: Journal of the Northern Plains* 65 (1998): 30.

Image 1.2. Religious practices, such as this First Holy Communion, *circa* 1938, reveal the process of cultural diffusion as the older generation teaches the youth about tradition, and the younger generation adapts to their new surroundings. Back row: Katherine Kautzman Doll; front row, left to right: Bertha Steckler, Thomas Wetch, and Melfrieda "Millie" Maryann (Doll) Hauck. *Photograph used with permission from the Melfrieda "Millie" Marie (Doll) Hauck Collection, GRHC.*

cultural diffusion allowed them (as well as other ethnic groups) to reconstruct their identity, thereby contributing to its resilience.

The German-Russian story began centuries ago, when thousands of people left the independent German *Länder* (states) for Russia. Germans from Russia "were land-hungry farmers who, in 1763, accepted the Enlightenment-influenced invitation of German-born Tsarina Catherine the Great of Russia to settle hitherto uncultivated land and to model modern agricultural techniques to surrounding peasants."[23] The tsarina even promised Germans autonomy to maintain their own identity. German immigrants, in fact, prospered for years in their secluded settlements along the Volga River, Black Sea, and other regions throughout Russia. Their lives, however, changed drastically once the Russian government reneged on Tsarina Catherine's promises and implemented a policy of Russification, a state-led effort to suppress minority cultures and enforce Russian language, customs, and military conscription. Around the same time, the mid1800s, Germans in Russia received information about free land and opportunities in America with the passing of the Homestead Act. Therefore, in the late 1800s, thousands of German-Russians left Russia. This double diaspora—the migration of Germans first to Russia and later to North America—forged a new, self-conscious ethnic group: German-Russians.[24]

The traditional narrative of the first generation was rooted in stories about never-ending trials and tribulations—the hard times in Germany, the persecutions in Russia, and the treacherous voyage to the United States. According to their story, once they were in the US or Canada, these hard times continued as thousands of Germans from Russia passed through ports of entry and migrated to the American or Canadian frontier. On this migration, they confronted unsavory swindlers and environmental calamities. Eventually, Germans from Russia settled on the Great Plains from Oklahoma to Saskatchewan. Volga Ger-

[23] Renate Bridenthal, "Reaching Out: The Double Diaspora of Germans from Russia" (paper presented at the *Heimat* Abroad: The Boundaries of Germanness, New York University, 1999).

[24] Bridenthal, "Germans from Russia: The Political Network of a Double Diaspora," in *The Heimat Abroad: The Boundaries of Germanness*, ed. Krista O'Donnell, Renate Bridenthal, and Nancy Reagin, *Social History, Popular Culture, and Politics in Germany* (Ann Arbor: University of Michigan Press, 2008), 187–218.

Image 1.3. Locations of Dakota Memories oral histories. *Map designed by William Clark.*

mans largely settled in the Central Plains, while Black Sea Germans primarily settled on the northern plains.[25]

According to their migration patterns, Black Sea Germans passed through South Dakota (usually Eureka or Yankton), and then migrated north. If the farmland proved unsatisfactory, they simply packed up and migrated again. In the early 1900s, for instance, thousands of Germans from Russia moved from the Sauerkraut Triangle (a small enclave of Germans from Russia in north-central South Dakota and south-central North Dakota) to western North Dakota and north-central North Dakota, creating the German-Russian Triangle (see

[25] Timothy J. Kloberdanz, "Volksdeutsche The Eastern European Germans: Hungry for Land, Hungry for a Home: North Dakota's 'Other Germans'," in *Plains Folk: North Dakota's Ethnic History* (Fargo, ND: North Dakota Institute for Regional Studies at North Dakota State University, 1988), 117-25; George Rath, "The Russian-Germans in the United States of North America, Preparations for Immigration," *Germans from Russia Heritage Society Heritage Review* 27, (September 1997): 15.

Image 1.3.). Some also continued their migration across the 49th parallel. Thus, Germans from Russia in Canada were largely the byproduct of a triple diaspora rather than a double diaspora.[26]

Germans from Russia began actively preserving their collective memory in the 1960s and 1970s as the first-generation immigrants passed away and grass-roots history and multiculturalism became increasingly popular. These internal and external dynamics led to the creation of historical societies and collections, such as the American Historical Society of Germans from Russia in Lincoln, Nebraska (1968); Germans from Russia Heritage Society in Bismarck, North Dakota (1971); and the Germans from Russia Heritage Collection in Fargo, North Dakota (1978). Focused on preserving the heritage of this ethnic group, these organizations were documenting its collective memory and identity.[27]

Most of the literature on Germans from Russia addresses stories from an adult perspective of Germans in Germany, Germans in Russia, or Germans in America. The works of Joseph Height and Karl Stumpp (two of the most recognized German-Russian scholars), for example, trace the immigration patterns of Germans in Europe, from the independent German states to the vast Russian steppe.[28] Scholars, such as Shirley Fischer Arends, George Rath, and Richard Sallet, have published landmark works addressing Germans from Russia in America, with discussions of German-Russian dialects, settlement patterns, and genealogy.[29] Timothy J. Kloberdanz and Ronald Vossler have also written

---

[26] For more information, see Bridenthal, "Germans from Russia," 187–218.

[27] "American Historical Society of Germans from Russia," www.ahsgr.org; "Germans from Russia Heritage Society," www.grhs.org; "Germans from Russia Heritage Collection," North Dakota State University, www.ndsu.edu/grhc.

[28] For more information, see Karl Stumpp, *The Emigration from Germany to Russia in the Years 1763 to 1862*, vol. 1 (Lincoln, NE: American Historical Society of Germans from Russia, 1973); Stumpp, *The Emigration from Germany to Russia in the Years 1763 to 1862*, vol. 2 (Lincoln, NE: American Historical Society of Germans from Russia, 1973); Joseph S. Height, *Paradise on the Steppe: The Odyssey of a Pioneering People*, 5th ed. (Bismarck, ND: Germans from Russia Historical Society, 1989); Height, *Homesteaders on the Steppe: Odyssey of a Pioneering People*, 2nd ed. (Bismarck, ND: Germans from Russia Heritage Society, 1987); Height, *Memories of the Black Sea Germans from Russia: Odyssey of a Pioneering People* (Bismarck, ND: Germans from Russia Historical Society, 1979).

[29] For more information, consult Shirley Fischer Arends, *The Central Dakota Germans: Their History, Language, and Culture* (Washington, D.C.: Georgetown University Press, 1989); George Rath, *The Black Sea German in the Dakotas* (Freeman, SD: Pine Hill Press, 1977); Rath, "The Russian-Germans in the United States of North America, Preparations for Immigration," 9–22; Richard Sallet, *Russian-German Settlements in the United States*, trans., LaVern Rippley and Armand Bauer (Fargo, ND: North Dakota Institute for Regional Studies, 1974).

extensively on the first generation, focusing on the continuation or evolution of folkways and traditions across several borders.[30]

Most of this historiography deals with the first-generation Germans from Russia and their ancestors, and it is mostly from the perspective of cultural insiders. In other words, Germans from Russia have primarily researched themselves, writing fascinating, self-affirming accounts of their history. Only a few academic historians or detached scholars, ones not themselves of German-Russian heritage, have published historical works addressing this ethnic group. Insider accounts are valuable to the historiography of Germans from Russia, as they add a unique perspective. Indeed, the insights of insiders, like Sallet and Kloberdanz, are essential.[31]

Nevertheless, there needs to be more examinations by outsiders, as well as more attention paid to the descendants of these immigrants, the children's story. Detached scholars or outsiders bring new perspectives to the discussion and reveal experiences or narratives that insiders may not, for one reason or another. Unfamiliar with the customs, outsiders may be more inquisitive in their interrogations and detailed in their explanations. Outsiders add new interpretations, not necessarily better but new nonetheless.

In addition, an examination of childhood history reveals a missing piece of the German-Russian story. In *Huck's Raft*, childhood historian Steven Mintz writes, "The history of childhood is inextricably bound up with the broader political and social events in the life of the nation . . . and children's experience embodies many of the key themes in American history."[32] Applying this on a micro level, the concept of childhood and the experiences of children are essential to the history of Germans from Russia on the northern plains. Therefore, the historiography of German-Russians is in need of not only an examination

---

[30] More information can be found in Kloberdanz, "Volksdeutsche The Eastern European Germans: Hungry for Land, Hungry for Home: North Dakota's 'Other Germans'"; Kloberdanz, "The Traditions We Share," 19–31; Kloberdanz, "The Tradition of Brauche: Facts and Fallacies," *Germans from Russia Heritage Review* 15 (September 1985): 3–5; Kloberdanz, "The Germans from Russia: A Viable Ethnic Group or Fading Phenomenon," *Journal of the American Historical Society of Germans from Russia* 27 (2004): 14–26; Ronald Vossler, *Dakota Kraut: A Memoir* (Fargo, ND: Germans from Russia Heritage Collection, 2003); Vossler, "Prairie *Spass*: The Folk Humor of North Dakota's Germans from Russia," *North Dakota History: Journal of the Northern Plains* 66 (Summer/Fall 1999): 50–60.

[31] These ideas are explored further in Clark, "A History Overlooked by Academic Historians: Germans from Russia on the Northern Plains" (paper presented at the Dakota Conference on Northern Plains History, Literature, Art, and Archaeology, Sioux Falls, SD, 22 April 2005).

[32] Mintz, *Huck's Raft*, viii.

Image 1.4. Herbert Herman fondly remembers his grandfather, Johann Roloff, who lived with his family during his childhood, *circa* early 1900s. *Photograph used with permission from the Herbert Herman Collection, GRHC.*

by outsiders, but also an investigation of German-Russian children (the second and third generations).

The field of childhood history traces its origins to the 1962 publication of *Centuries of Childhood* by cultural historian Philippe Ariès, a work that sparked ongoing scholarly debates on the historical role of children and the concept of childhood.[33] In this foundational work, Ariès examines the evolution of childhood, arguing that children before the modern era never experienced child-

[33] Cunningham, *Children and Childhood*, 4.

hood.[34] More importantly, Ariès draws attention to the changing concept of childhood, and to the differences between the terms children and childhood. Children, as Ariès concludes, refers to an age group, while childhood is a social construct imagined and defined by a group of individuals.[35]

For years, scholars largely overlooked Ariès's work. In fact, there was no response to his publication for ten years after publication. "When social history began to enjoy a boom in the late 1960s and early 1970s," Hugh Cunningham points out, "*Centuries of Childhood* was the only available text on its topic."[36] By the mid-1970s, however, scholars began turning their attention to the history of children and childhood in Europe, thereby directly responding to the work of Ariès.[37]

Academic historians of the United States largely overlooked these topics until the late 1980s. Then, in 1989, Western American historian Elliott West published *Growing up with the Country*, thus becoming the premiere historian of childhood in the American West. In this account of growing up on the American frontier, West insists children have agency. Acknowledging that some may question his rationality, West writes, "for three reasons in particular, the children's history stands on its own."[38] First, "children perceive the world differently than adults do."[39] Second, children had their own distinct role in settling the frontier. They were active in migrating west, establishing farmsteads, and maintaining farms. Third, "the children's story adds something to western history's most familiar theme – the ways in which the frontier changed those who lived there and thus created a new and distinctive part of America."[40]

Two years after West's optimistic work, an English professor from the University of North Dakota published *Settlers' Children*, a pessimistic work on growing up on the Great Plains. Rather than establishing children as historical actors in the West, Hampsten perceives them as victims. She insists that children and women were sufferers of cruel fathers and husbands, who made them

---

[34] Philippe Ariés, *Centuries of Childhood: A Social History of Family Life* (New York: Random House, 1962), 33.

[35] Ariés, *Centuries of Childhood*, 9–10.

[36] Cunningham, *Children and Childhood*, 7.

[37] For more information, see Lloyd de Mause, *The History of Childhood* (New York: Psychohistory Press, 1974); Edward Shorter, *The Making of the Modern Family* (London: Collins, 1976); Lawrence Stone, *Family, Sex, and Marriage in England 1500–1800* (New York: Harper and Row, 1977).

[38] West, *Growing Up with the Country*, xviii.

[39] West, *Growing Up with the Country*, xviii.

[40] West, *Growing Up with the Country*, xviii.

venture out into the harsh unknowns of the Great Plains. She argues that men were borderline abusive for taking their families on this dangerous and unnecessary journey west. Her interpretation portrays children as weak, malleable victims rather than young, excited adventurers.[41]

The works following West's *Growing up with the Country* and Hampsten's *Settlers' Children* provide a more even-handed analysis, discussing the trials and fortunes of growing up in the American West. In *Children of the Western Plains*, historian Marilyn Irvin Holt provides a cursory glance of immigrant and minority children. Holt, like most childhood historians, takes a topical approach, with chapters devoted to migration, family, work, school, and play. She provides thorough discussions on each topic and concludes that growing up in the West had both highs and lows. Holt, however, groups all immigrants together, simplifying their narratives.[42]

In 2005, rural historian Pamela Riney-Kehrberg published one of (if not the) most thorough accounts of children and childhood in rural America. In *Childhood on the Farm*, Riney-Kehrberg sets out to debunk some historical myths, as well as to address three spheres of influence on rural childhood: work, school, and play. Nevertheless, she declines to discuss the role of ethnicity in the lives of farm children. She says, "[W]here ethnicity made a significant difference in the fabric of children's lives, such as the work roles of girls, I have noted these distinctions."[43] Yet, this occasional reference to ethnicity downplays its role in rural childhood. Her decision to overlook ethnicity may correspond to another significant oversight—religion as a sphere of influence. Although Riney-Kehrberg argues that work, school, and play were the three primary forces in the lives of children, for descendants of immigrants, ethnicity and religion were equally powerful forces.[44]

Thus, since West's 1989 publication *Growing up with the Country*, scholars of American history have joined those of European history in turning their attention to an often-overlooked group: children. West, Holt, and Riney-Kehrberg have made strides in explicating the roles of children, as well as the concept of childhood, in the American West. They discuss a variety of topics: migration, settlement, relationships, perceptions, work, school, and play. Then again, they

---

[41] For more information, see Hampsten, *Settlers' Children*.

[42] See Holt, *Children of the Western Plains*.

[43] Riney-Kehrberg, *Childhood on the Farm*, 4.

[44] Riney-Kehrberg, *Childhood on the Farm*, 4.

Image 1.5. John Riehl's brothers (left to right) Kasper and Martin going for a ride in their little wagon, 1933, a simple joy that captures the playfulness and strong sibling bonds among German-Russian children on the plains. *Photograph from the John and Ann (Gray) Riehl Collection, GRHC.*

(as well as other historians of childhood in the American West) largely neglect ethnic diversity, and primarily focus on children in the nineteenth century. *Echoes of the Old Country* adds to this existing historiography by examining ethnic children, such as German-Russians, in the twentieth century, "the century of the child."[45]

Dubbed the *child's century,* from 1870 to 1960 the social construction of childhood evolved, creating an era of people fixated with the physical, mental, and emotional well-being of youth in the western world. Historians like Hugh Cunningham attribute this new focus on children to the reformers of the Progressive Era, who desired to assimilate immigrants into mainstream American society, thus reducing the language and cultural diversity in America. While most Progressive initiatives unfolded in urban and industrialized centers, the

[45] Mintz, *Huck's Raft*, 373.

ideals of this movement migrated across the globe. And, although the ideals were not always the reality, families in the twentieth century, for the most part, began to see children as "economically worthless" and "emotionally priceless."[46]

By 1917, many of the Progressive reforms and ideals fell by the wayside, as American adults became increasingly preoccupied with the realities of World War I. Yet, societies throughout western civilization, including in America, continued to transform the definition of childhood. For instance, at this time, the social expectations in many industrialized societies dictated that children ages five to fourteen receive a formal education. While this ideal was not always feasible for all social classes in every region of the nation (especially in rural communities), it was nonetheless the social expectation. By 1918, all states in America had established compulsory education laws, as well as other educational mandates. As sociologist Viviana Zelizer writes, the "sentimentalization of childhood intensified regardless of social class" throughout the early 1900s.[47] Concerned with protecting and shielding the youth from the burdens of adulthood, American adults prioritized the wellbeing of children.

That being said, with the onset of the Great Depression in the late 1920s and early 1930s many communities in the United States witnessed a temporary setback. The societal and familial need for child labor during these hard times made it improbable, if not impossible, for children to be "economically worthless."[48] While the compulsory education laws mandated children attend a certain number of days of school each year—the specific requirements of educational laws varied state by state, not the federal government—the necessity for additional family income or farm labor outweighed the consequences for truancy. Thus, many families opted not to send their children to school, and instead sent their children to the fields or factories.

The outbreak of World War II in 1939 and the US involvement by 1941, however, shifted the focus back onto the Progressive ideals—a protected, secure, and long childhood for all. With men fighting in the war and women working in the war industries, increasingly children could be found in school—a safe place that provided care (including meals with the new state-funded lunch programs) and education. According to Cunningham, a childhood historian,

---

[46] For more information, see Cunningham, *Children and Childhood*; Viviana A. Zelizer, *Pricing the Priceless Child: The Changing Social Value of Children* (Princeton: Princeton University Press, 1985).

[47] Zelizer, *Pricing the Priceless Child*, 209.

[48] Zelizer, *Pricing the Priceless Child*, 209.

"school, not the workplaces was the proper place for children to be when [their parents] were not at home."[49]

By the end of World War II, the so-called century of the child was fading fast.[50] A new era was on the horizon—one with television, consumerism, and commercialization—redefining the social construction of childhood. There was a whole new level of complexity, by the 1960s, as the concept of adolescence emerged—separating teenagers (ages thirteen to nineteen, or by marriage) from both the constructions of childhood and adulthood, placing them in a liminal stage of life. On the one hand, adults wanted to continue prolonging dependency, a continuation from the Progressive Era, while teenagers began demanding more freedom and opportunity for fun. This conflicting period of generational tension combined with an advertising industry reaching out to children through television, allowed the place of children in America to become more tumultuous.[51]

All that being said, during the first half of the twentieth century, that is, before 1960, most "children were insulated from the adult world."[52] While aware of the world events, like World War I, the Great Depression, and World War II, most children were preoccupied with being children rather than the hard times and realities experienced by adults. After all, social norms during the period from 1870 to 1960 emphasized childhood as a time of happiness, security, and prolonged innocence.[53]

Using the oral histories and archival documents of the Dakota Memories, *Echoes of the Old Country* reveals the ethnic childhood of the second- and third-generation German-Russians on the northern plains during a child-centered century, at least for those who participated in this oral history project. Determined to maintain the stories of their ancestors, these German-Russian descendants ground their narratives in their ethnic group's collective memory, a memory created long before the first-generation Germans from Russia touched American or Canadian soil. They hold onto the traditional story of oppression and hardship, yet the Dakota Memories narrators reveal layers of complexity through their storytelling. While these descendants root the German-Russian story in their ancestral past, their childhood experiences and present conditions

[49] Cunningham, *Children and Childhood,* 181.
[50] Mintz, *Huck's Raft*, 373.
[51] Cunningham, *Children and Childhood*, 187.
[52] Cunningham, *Children and Childhood*, 189.
[53] Mintz, *Huck's Raft*, 373.

also play a role in shaping it. That is to say, the story of German-Russians is not static. By sharing their memories with oral historians, these German-Russian memory-makers have created a new story, a story of optimism and opportunity. The ability of these narrators to bridge their distant past, recent past, and present contributes to their success at maintaining their ethnic identity.

One such example is Ben Ferner, of Katheriental, Saskatchewan, whose tragic yet inspirational story highlights the adaptability and resilience of his community. As Ben tells it, his childhood would have ended before it started without the support of both his father and the people around him. When he was just fourteen months old, Ben's mother passed away, leaving behind her husband and child. Though family members offered to take him in, his father, George Ferner, could not bear to part with his only son. Instead of raising Ben alone, he relied on the generosity of neighbors, who helped care for the boy while he worked on a nearby farm. During the day, Ben ate, played, and learned in their homes before returning to his father each evening. This deep tradition of community support ensured that no one faced hardship in isolation. For generations, neighbors stepped in to assist with childcare, farm work, and financial burdens, strengthening bonds of mutual aid. Ben's story encapsulates the dual narrative at the heart of this history—one of hardship and loss, but also of resilience and collective strength. Individual struggles were met with communal care, reinforcing a legacy of perseverance that defined their way of life.[54]

---

[54] Interview with Ben Fermer; Jessica Clark and Thomas D. Isern, "Germans from Russia in Saskatchewan: An Oral History," *American Review of Canadian Studies* (March 2010), 77.

CHAPTER 2

Amber Waves: The Northern Plains Environment

Delmar D. Zimmerman, from Wishek, North Dakota, truly enjoyed his rural childhood, and claims it was unforgettable. He has fond memories of the northern plains environment, including the picturesque sunrises and sunsets. When asked about his memories of growing up on the northern plains, Delmar mentions the beautiful flowers and the amazing fields with "amber waves of grain." He is not oblivious to the trials and tribulations of the environment, however, as he also reminisces about the tough winters and scorching summers. Although a young boy during the 1930s, Delmar cannot forget the severe droughts with blowing dirt and thistles. He insists the thirties were some of his most unpleasant years. Nevertheless, Delmar has a deep appreciation for his childhood on the Plains—not because it was without hardship, but because it was home. Life was shaped by the land, the seasons, and the strength of family and community, and even the most difficult moments were simply part of what it meant to grow up there. Looking back, he does not dwell on the struggle, but rather on the deep sense of place and identity that the plains gave.[1]

For the second- and third-generation German-Russians, like Delmar, the northern plains environment played a vital role in shaping their childhood, as well as their identity. This environment not only contributed to their most trying and terrifying times, but also gave them opportunities for fun and enjoyment. Children experienced a number of environmental calamities, such as prairie fires, dust storms, grasshopper invasions, blizzards, and extreme weath-

[1] Interview with Delmar D. Zimmerman.

er changes. The environment also allowed children to roam fields, climb trees, pick berries, hunt deer, and trap gophers. Memories of this environment encouraged the descendants of Germans from Russia to maintain and strengthen their identity, building an identity of survivors as well as people of the plains. Unlike their grandparents and parents, American-born and Canadian-born German-Russian children had no basis for an environmental comparison. They were unfamiliar with the Russian steppe of their parents, as they only knew the northern plains. In the words of historian Elliott West, they were "growing up with the country."[2] Thus, during their oral histories for Dakota Memories, narrators reconstructed their ethnic story to accommodate these childhood experiences with the terrifying and the tranquil aspects of the northern plains environment.

In the mid-1800s, Germans from Russia fleeing the Black Sea and Volga River regions of Europe primarily settled in the Great Plains or Great American Desert of North America. The Great Plains span from the Canadian provinces of Alberta, Saskatchewan, and Manitoba to the American state of Texas (see Image 2.1). It is a distinct area shaped by alluvial deposits, where sediment carried down from the mountains by meandering rivers accumulates over time. These deposits form a broad, sloping apron of material that spreads across the landscape. However, due to continuous erosion and shifting waterways, the topography of the plains is in a constant state of change. According to William C. Sherman—who located and mapped ethnic settlement across North Dakota—however, the region is "flat and featureless."[3] The leading Great Plains historian, Walter Prescott Webb, uses three words to characterize this steppe region, "level," "treeless," and "semi-arid."[4]

Within the Great Plains, the northern plains is a sub-region that has distinct characteristics. The so-called Sauerkraut Triangle, the core region inhabited by predominately Black Sea Germans, hardly resembles the descriptions of Sherman and Webb. This grassland area is part of the Missouri Coteau region of rolling hills, grasslands, and wetlands. This region's characteristics resulted from a glacial melt that occurred nearly ten thousand years ago and left behind valleys and glacial deposits. Hence, the land consists of rolling grasslands,

---

[2] For more information, see West, *Growing Up with the Country.*

[3] Walter Prescott Webb, *The Great Plains* (Boston: Ginn & Company, 1931), 3–11; William C. Sherman, et al., *Plains Folk: North Dakota's Ethnic History* (Fargo, ND: North Dakota Institute for Regional Studies at North Dakota State University, 1988), 3.

[4] Webb, *The Great Plains*, 3–16.

Image 2.1. The Great Plains. The northern plains consist of South Dakota and North Dakota, as well as parts of Montana, Saskatchewan, and Manitoba. *Map permission from Professor Thomas D. Isern co-author of Plains Folk website, www.plainsfolk.com. Original drawing by geographer Bill Phillips of Emporia State University. Updated by Deb Tanner, NDSU Extension.*

mixed-grass prairie, and wetlands of various sizes. The grasslands of this region are unparalleled. According to the website for Chase Lake National Wildlife Refuge, this area consists of "a mixture of short and mid-grasses dominated by blue gamma and green needlegrass, with many species of wildflowers."[5] It is also an area with distinct berries and edible plants, such as juneberries and chokecherries. In other words, this northern region of the Great Plains is anything but "flat and featureless."[6]

In addition to a distinctive topography, the climate in this region was then, as it is today, varied and unpredictable. The northern plains is a semi-arid re-

[5] "Wetlands," United States Fish and Wildlife, https://www.fws.gov/refuge/chase-lake html (15 March 2025).
[6] Sherman, et al., *Plains Folk,* 3.

gion. In his foundational work on North Dakota history, historian Elwyn B. Robinson acknowledges that "semi-aridity makes North Dakota a borderland between the humid East and the arid West."[7] This statement is representative for all states and provinces located in the northern plains region. The annual precipitation in the state of North Dakota ranges from less than thirteen inches in the western regions to more than twenty inches in the eastern regions, which is significantly lower than the Midwest or eastern United States, where annual precipitation often exceeds forty inches. The snowfall in this state is also considered moderate compared to regions like the Great Lakes or the Northeast—ranging from roughly twenty-six inches in the west to approximately thirty-eight inches in the east. While snowfall may seem substantial, it contributes less to overall moisture because of high evaporation rates and dry winds. The lack of precipitation in North Dakota is representative of the northern plains region as a whole, where moisture levels are often insufficient for consistent crop yields without irrigation.[8]

People living in this region confront a variety of climatic extremes, such as strong winds, tornados, blizzards, droughts, and dust storms. In *The Great Plains*, Webb states, "[T]he wind blows harder and more constantly on the Plains than it does in any other portion of the United States, save on the seashore."[9] During the winter months, people struggle with drifting snow, subzero temperatures, and excessive wind chills. At times, the snow is as high as the houses and outbuildings, and in the cracks and crevices of everything. For months, it is too cold for snow to melt. In the summer months, windstorms, thunderstorms, and sandstorms force people to adapt or change accordingly. As Webb observed, survival on the plains required ingenuity, leading to innovations such as sod houses, which provided insulation against extreme temperatures, and windmills, which helped harness the relentless winds for water and energy. Farmers adjusted their agricultural practices by using drought-resistant crops and deep-plowing techniques to retain soil moisture, while ranchers learned to manage livestock with windbreaks and shelterbelts[10] The fluctuating temperature on the northern plains contributes to the region's environmental

---

[7] Elwyn Robinson, *History of North Dakota* (Lincoln, University of Nebraska Press, 1966), 8.
[8] "Climate of North Dakota," US Department of Interior," https://gf.nd.gov/wildlife/habitats/climatehtm (15 March 2025).
[9] Webb, *The Great Plains*, 21.
[10] Webb, *The Great Plains*, 21.

distinctiveness. Sherman writes, "temperatures fluctuate from oppressive summer days to frigid artic moments."[11] For instance, the record high for North Dakota and South Dakota were set in 1936: 121°F and 120°F respectively. The record lows for both states were also set that year: -60°F in North Dakota and -58°F in South Dakota.[12] Historian Elliott West accurately asserts, "the weather of the northern plains was some of the most erratic in the world."[13]

Complementing this distinct geography and climate, the seasons are unique compared to other areas in North America and the season lengths are varied. Spring is typically the shortest season in the region, lasting between one to two months, usually April and May. Wetness characterizes this season and is a direct result of melting snow. Thus, it is common for parts of this region to experience spring flooding. Summer in the region is the second longest season, June through August. It consists of long, warm days and short, cool nights. Thunderstorms are frequent and sometimes quite severe, occasionally producing tornadoes. Fall lasts for two to three months, September through November, with warm days, cool nights, and light rain. The last season—and the region's longest—is winter, typically lasting between five to six months, from November to April. The cold is often extreme: subzero temperatures, dangerous wind chills, and intense blizzards.[14]

The nature of this environment has stimulated the region's biodiversity. A wide range of species, including ungulates, carnivores, birds, rodents, reptiles, amphibians, fish, and insects, thrive in this ecosystem. Among the common hoofed animals residing on the northern plains are deer and antelope, both of which are well adapted to the harsh conditions. The region also supports a diverse population of predators, including bears, wolves, coyotes, skunks, and wildcats. Additionally, with the vast grasslands and plentiful wetlands, this region is an ideal breeding ground for birds, such as passerines, shore birds, and raptors. Lagomorphs and rodents, including jackrabbits and gophers, are also native to the region, having inhabited these plains for thousands of years.[15]

---

[11] Sherman, et al., *Plains Folk*, 3.

[12] "National Weather Service," United States Department of Commerce," www.weather.gov (21 November 2009).

[13] West, *Growing Up with the Country*, 4–5.

[14] Robinson, *History of North Dakota*, 8–11.

[15] For more information, see Robert Seabloom, *The Mammals of North Dakota*, 2nd ed. (Fargo, ND: North Dakota State University Press, 2020).

Abundant animals are a significant component of the region's ecological identity and environmental history.[16]

For years, memoirists from the northern plains have written extensively about their recollections of this environment. In fact, the environment usually dominates accounts of growing up in this region. For instance, German-Russian memoirists Pauline Neher Diede and Debra Marquart devote considerable space to descriptions of the environment in *Homesteading on the Knife River* and *The Horizontal World*, respectively.[17] Other regional memoirists outside the German-Russian ethnic group write extensively on their interactions with the environment, as well. For instance, in Wallace Stegner's *Wolf Willow: A History, a Story, and a Memoir of the Last Plains Frontier*, Lois Phillips Hudson's *Reapers of the Dust*, and Dorothy Hubbard Schwieder's *Growing up with the Town*, the environment clearly plays an important role in child development.[18]

In *Wolf Willow*, a fictionalized memoir depicting life in Saskatchewan during the early 1900s, Stegner emphasizes the role of the environment in shaping childhood. According to him, "Saskatchewan can be a pretty depressing country."[19] He describes the emptiness of the landscape and the violent and prolonged weather patterns. However, he also acknowledges its deeper significance, writing, "viewed personally and historically, that almost featureless prairie glows with more color than it reveals to the appalled and misdirected tourist."[20] Stegner presents a complex relationship with the land, describing it as both harsh and unforgiving yet also romantic, dignified, and deeply beautiful. He places the environment at the center of his work on history and memory.[21]

Hudson also focuses on the role of the environment in her collection of fictional short stories, *Reapers of the Dust*. Like Stegner's *Wolf Willow*, Hudson's work is rooted in personal experiences, yet written as a fictional narrative.

---

[16] John L. Allen, "Landscape Change at the Confluence: From Lewis and Clark to the Present," *North Dakota History: Journal of the Northern Plains* 69 (2002): 2–23; "Wildlife and Conservations," *North Dakota Be Legendary*, https://gf.nd.gov/wildlife (15 March 2025).

[17] For more information, see Diede, *Homesteading on the Knife River Prairies*; Debra Marquart, *The Horizontal World: Growing up Wild in the Middle of Nowhere* (New York: Counterpoint, 2006).

[18] For more information, see Wallace Stegner, *Wolf Willow: A History, A Story, and A Memory of the Last Plains Frontier* (New York: Penguin Books, 2000); Lois Phillips Hudson, *Reapers of the Dust: A Prairie Chronicle* (Minneapolis, MN: Minnesota Historical Society Press, 1984); Dorothy Hubbard Schwieder, *Growing up with the Town: Family and Community on the Great Plains* (Iowa City, IA: University of Iowa Press, 20002).

[19] Stegner, *Wolf Willow*, 5.

[20] Stegner, *Wolf Willow*, 4.

[21] For more information, see Stegner, *Wolf Willow*.

She sets this story in the central region of North Dakota during the 1930s, describing the region's dust storms, blizzards, and excessive temperatures. Hudson, however, never lets these environmental calamities stop her main character from exploring the countryside, trapping gophers, and catching frogs. She portrays her character as a young, adventurous girl, who overcomes hardships and triumphs.[22]

In a nonfiction memoir, Schwieder recounts her childhood as a third-generation immigrant in Presho, South Dakota. As in the other memoirs, the environment is prevalent throughout her work, *Growing up with the Town*. She reminisces about the hardships of life on the plains, including prairie fires, scarce water, and brutal winters. She describes the simple structure her family built upon arriving in Dakota Territory—a sod house. Nevertheless, Schwieder, who grew up during the 1930s and 1940s, recalls finding joy in the landscape that shaped her childhood. She writes about summer days spent fishing and swimming in local creeks, exploring the open prairie, and making games out of the land's natural features. Her memoir describes the environment as an influential, yet not victimizing, force. Despite its challenges, the plains provided her with freedom, adventure, and a deep connection to the land.[23]

Dakota Memories narrators speak at length about their memories of the northern plains environment. They share stories and perceptions of life during each season, and some discuss the peace and serenity of childhood in a rural setting while others reminisce about the hardships they had to endure. Regardless, it is apparent that the environment had a direct impact on the social, cultural, and physical development of the second- and third-generation German-Russians. They grew up interacting and dealing with this region's rigorous environment.

Spring seems to hold a special place in the memories of several narrators. There was something peaceful and surreal about living in the wide-open country during this season. With the changing weather conditions, such as increasing temperatures, children were able to enjoy the vastness of the region. They found comfort in exploring their surroundings, given their familiarity with the landscape and environment. As most Dakota Memories narrators insist, the northern plains was all they had ever known. German-Russian descendants found pleasure in the sounds and smells that accompanied spring season, like

[22] For more information, see Hudson's *Reapers of the Dust*.

[23] For more information, see Schwieder, *Growing up with the Town*.

the "whistling wind and blowing dirt."[24] Of course, freshly turned earth and newly blossomed lilacs trigger sweet memories as well. The sights, sounds, and smells children encountered with the onset of spring were nothing short of "refreshing," as they represented the end of the long, cold winter months.[25]

Spring season also brought about a rejuvenation of life for both nature and people. As the migratory birds and mammals began trekking back into the region, plains people began preparing fields and gardens. For some children, it was energizing to hear the firing up of tractors and farm machinery after a long winter.[26] Marvin L. Hartmann of Delmont, South Dakota, for instance, insists these were some of the best sounds in the world. He really enjoyed growing up in the open country of South Dakota. Thinking back, Marvin describes the aroma of the freshly turned earth as simply fantastic. He also reminisces about the smells of blossoming lilacs and alfalfa fields. For him springtime was refreshing.[27]

For a few Dakota Memories narrators, spring was the best season of the year, as increased opportunities for fun and enjoyment accompanied it. For school-aged children, spring meant more time outside during noon-hour and recess, and the opportunity to play more outdoor games, like ball and Anti-I-Over. Ball was an informal version of baseball. Students looked forward to playing ball despite rarely having the proper equipment—gloves, balls, and bats. Many times, children used their natural surroundings to make their own equipment, such as tree branches or fenceposts for bats and rocks or socks for balls. Another favorite game, Anti-I-Over, was played by dividing into two teams and standing on opposite sides of a small building, often a schoolhouse or barn. One team would throw a ball over the roof while shouting "Anti-I-Over!" If the opposing team caught the ball, they would run around the building and try to tag members of the other team before they could reach the opposite side, switching roles in the process. The game was fast-paced, simple, and required nothing more than a ball and an available rooftop, making it a favorite among children on the plains. According to Dakota Memories narrators, they cared

---

[24] Interview with Marvin L. Hartmann.

[25] Ed. note: Each interviewee's name appears in full at its first reference. In subsequent references, only the last name appears (unless multiple interviewees share the same last name). Interviews with Hartmann, Alice Ruth (Miller) Buck, ElRoy Fischer, Herbert Herman, Michael Jahner, Rupert Kleingartner, Philip Lang, Agatha "Peggy" Sellinger, and Deloris (Boschee) Zimmerman.

[26] Interviews with Alice Buck, ElRoy Fischer, Hartmann, Rupert Kleingartner, Philip Lang, and Deloris Zimmerman.

[27] Interview with Hartmann.

Image 2.2. Christina Rose Marie (Bast) Krismer with her brother Gabriel and sister Bernice playing in their mother's garden, *circa* 1945, illustrating how German-Russian children turned work spaces into places of joy and learning. *Photograph from the Christina Rose Marie (Bast) Krismer Collection, GRHC.*

less about the type of games or equipment, simply enjoying the opportunities for more outdoor activities.[28]

Children not only engaged in these types of activities at school during the spring season, but also at home. Despite having an increased workload this time of year, children found time for simple pleasures, such as making mud pies and playing make-believe.[29] Alice Ruth (Miller) Buck, a native of Streeter, North Dakota, born in 1932, has fond memories of making mud pies. She spent much of her youth playing with her older sister, Leona, and her younger brother, Norman. As little girls, Alice and Leona often crept into their barn to steal sparrow

[28] Interviews with Alice Buck, Hartmann, Herman, and Jahner.
[29] Interviews with Alice Buck, Hartmann, and Herman.

eggs from the nesting birds. They used the eggs for their secret recipe for mud pies. [30] Another favorite pastime of prairie children was playing make-believe with miniature farm equipment in their mothers' gardens. Children enjoyed mimicking the behaviors of their parents and older siblings preparing the fields and gardens for spring seeding.[31]

During their childhoods, narrators also found enjoyment in roaming the revitalized land. The spring foliage was spectacular, with the flowers blossoming, the trees budding, and the plants sprouting. Thus, a favorite pastime was simply roaming the fields and picking flowers. Lilacs, prickly pears (a yellow flower that produces a fig-like fruit), and prairie crocuses (a bluish-purple prairie flower) were among the commonly picked flowers.[32] Furthermore, children spent endless hours in the fields with their fathers and mothers. As their parents were tilling the land and planting seeds, children wandered around engrossed by their surroundings.[33]

Trapping was by far one of the most popular spring pastimes for the second- and third-generation German-Russians interviewed for the Dakota Memories project.[34] Some narrators recall trapping for pure enjoyment, while others recall trapping for spending money. For many, trapping was a practical skill passed down from older family members, providing a way to make extra income through the sale of pelts, which were valued in the fur trade. Trapping was a common practice on the northern plains because lagomorphs and rodents were a nuisance for farmers. Plains children found this activity invigorating, as well as peaceful. It involved so much more than trapping and killing the pesky varmints. First, children usually spent time scouting the pastures and fields to find the best locations to set the traps. They often targeted gophers, muskrats, and weasels, using simple box or steel traps, carefully placing them in areas where they had observed animal activity. Then, after setting up the traps, it was a waiting game. For a few Dakota Memories narrators, like Herbert Herman, from Gackle, North Dakota, waiting was the best part. Herbert recalls spending

[30] Interview with Alice Buck.

[31] Interviews with Pauline "Polly" (Wock) Berger, Alice Buck, and Deloris Zimmerman.

[32] "Native Wildflowers of North Dakota Grasslands: Common Pincushion Cactus (*Coryphantha vivipara*)," United States Department of Interior, www.npwrc.usgs.gov/resource/plants/wildflwr/species/coryvivi (21 November 2009); "Prairie Wildflowers and Grasses of North Dakota: Spring Flowers," United States Department of Interior, www.npwrc.usgs.gov/resource/plants/pwgofnd/prairc (21 November 2009).

[33] Interviews with Alice Buck, ElRoy Fischer, Hartmann, Herman, Jahner, and Deloris Zimmerman.

[34] Esther (Haerter) Drefs interview; Hartmann interview; Herman interview.

Image 2.3. Walter Kattlerling and his brother Harold digging out the cows in the barn to milk them, 1932, showing the harsh prairie winters and the determination of German-Russian families to keep daily life going no matter the conditions. *Photograph from the Ellean M. (Burack) Kattlerling Collection, GRHC.*

days simply lying on his back watching the cloud formations and waiting for his traps to snag a rodent. He, as well as other people of the plains, remembers feeling alive and free out in the wide-open country.[35] Beyond recreation and profit, trapping also served a functional purpose in controlling pest populations, particularly gophers, which could damage crops and pastureland. When a trap snagged a rodent, children killed it, chopped off the tail, and turned the tail in for a few cents. It was a pastime activity as well as a part-time job.[36]

Spring was not without its difficulties. Growing up during the drought years of the 1920s and 1930s, Dakota Memories narrators reminisce about the spring blizzards and dust storms.[37] By late March or early April, people were ready for winter to end and the snow to melt, but blizzards often swept through the plains well into mid or late April, making headlines in the local newspapers.

---

[35] Interview with Herman.

[36] Interviews with Herman, interview; Esther (Haerter) Drefs, and Jacob A. Helbling.

[37] Interviews with Polly Berger, Hartmann, Jahner, Rupert Kleingartner, interview; Julius Miller, Arthur E. Schmidt, and interview; and Peggy Sellinger.

On April 9, 1937, the *Napoleon Homestead* ran a story announcing "10 Inches of Snow Boon to Seeding." According to the article, farmers had mixed reactions—some welcomed the moisture while others saw it as a frustrating delay to seeding during an already difficult drought.[38]

The late snow blizzards and early dust storms created additional work for families on the northern plains. They would have to use shovels to dig out their homes, barns, fences, and farm equipment.[39]

In addition to coping with the unpredictable spring weather, children also dealt with increasing workloads and responsibilities this time of year. Some German-Russian descendants insist there was nothing beautiful or enjoyable about this season.[40] Rupert Kleingartner from Gackle, North Dakota, recalls his childhood as largely dull and boring, with never-ending chores. According to him, growing up German Russian on the northern plains was all about work—"work from sun-up to sun-down." Spring was especially hard, Rupert insists, as his family usually began making a fuel source for the upcoming winter using dried manure, commonly referred to as *mischt* among German-Russians. It was a long, smelly, and tedious four-step process. Step one consisted of gathering the manure in a relatively large, flat area. The second step involved running horses around in circles on the manure until it became hard. Step three entailed cutting the hardened manure into blocks small enough for the kitchen stove. Finally, the last step involved moving and stacking the dried mischt in a central location to be used as fuel. Rupert never enjoyed this process, as it was monotonous and foul smelling.[41]

Making mischt blocks was not the only unpleasant seasonal chore that accompanied spring. Narrators also recall the tedious and smelly chore of hauling away the winter insulation.[42] It was common practice in the early 1900s to push manure against the houses in the winter months. People usually packed it to a depth of two to three feet around the foundation. Some, like Arthur E. Schmidt's Neudorf Township, North Dakota, family sealed their foundation

---

[38] "10 inches of Snow Boon to Seeding: Snowfall General in State; Moisture Totals 4.50 In. Since January 1," *Napoleon Homestead*, 9 April 1937.

[39] Interviews with Hartmann, Rupert Kleingartner,; Joseph George Kohlman, interview; Julius Miller, interview; Agatha "Peggy" Sellinger, interview.

[40] Rupert Kleingartner interview; Joseph George Kohlman, Julius Miller, and Schmidt.

[41] Interview with Kleingartner.

[42] Interviews with Kleingartner, Kohlman, and Schmidt.

with cardboard before packing it with manure; others, like Joseph George Kohlman's family from Tramping Lake, Saskatchewan, simply pushed the manure right up against the foundation. The manure provided added insulation and heat during the cold winter months. There was a small drawback, however, for as Arthur and Joseph attest, it made for an unpleasant chore at the onset of spring. As the temperatures rose and the snow melted, the manure began to smell. Accordingly, the insulation needed to be moved to the pastures and fields as soon as possible.[43]

Field and garden preparation also increased the workload for rural children.[44] With the onset of spring, farmers began priming their equipment for seeding. Young boys drove horses or tractors to help till the land; following tilling, children also helped to sow the grain. Because German-Russian families had extensive gardens before the outbreak of World War II, narrators recall helping their parents plant and cultivate. For the most part, labor was gender divided—men and boys worked in the fields while women and girls took care of the house and garden. Nonetheless, girls helped with the fieldwork, and a few boys assisted with some domestic chores.[45]

Memories and stories of spring on the northern plains reaffirm the German-Russian story. According to narrators, children survived spring by coping with and relying on their environment. Springtime provided children opportunities, as well as hardships. The narrators remember finding ways to interact with the environment for pleasure, such as wandering fields and trapping gophers. They also used the natural environment to cope with climate hardships, like using manure for fuel and insulation. This practice of using dried mischt for fuel was an adaptation carried over from their ancestors in their German-Russian old country, where harsh winters and limited access to wood forced settlers to find alternative heat sources. Similarly, reliance on communal labor and resourcefulness in facing extreme weather mirrored the survival strategies their ancestors had developed on the Russian steppe. Dakota Memories narrators, in other words, use their memories of trying times to maintain their ethnic group's traditional narrative—one rooted in overcoming environmental hardship, displacement, and self-sufficiency—and their memories of enjoyment to personalize it. By weaving together memories of struggle and resilience—both

[43] Schmidt interview; Kohlman interview.

[44] Interviews with Polly Berger, Herman, and Deloris Zimmerman.

[45] Interviews with Polly Berger, Herman, and Deloris Zimmerman.

Image 2.4. Richard Kraft with a lamb and Marlene Ann (Kraft) Smith holding a kitten, reflecting how caring for animals taught German-Russian children responsibility, compassion, and connection to farm life. *Photograph from the Marlene Ann (Kraft) Smith Collection, GRHC.*

inherited and experienced firsthand—they preserve and adapt their ethnic identity in a way that continues to thrive.[46]

Unlike spring, summer is a season the majority of narrators of the Dakota Memories discuss at length. Some emphasize the pleasant memories of pastimes, like fishing and swimming, while others focus on the unpleasant memories of environmental hardships, such as prairie fires and grasshopper invasions. Regardless, the second- and third-generation German-Russians have clear and vivid memories of summers on the northern plains. Their experiences and memories of this season, as well as the others, have clearly allowed them to maintain and reshape their ethnic group's narrative—a narrative created by the first generation. As the original settlers attested, Germans from Russia were plains people who faced obstacles and persevered. Their descendants, however,

[46] Interviews with Alice Buck, Polly Berger, ElRoy Fischer, Hartmann, Herman, Kohlman, Philip Lang, Julius Miller, Schmidt, Peggy Sellinger, and Deloris Zimmerman.

reminisce about more than the environmental hardships. They remember the opportunities, as well.[47]

A number of German-Russian descendants label summer as the best time of year on the northern plains.[48] The wide-open prairie on warm, long days was nothing short of spectacular. The region's level landscape unhindered by an abundance of tall objects, like trees and mountains, provided inhabitants with the ability to see for a great distance. The wide-open space comforted prairie people. Dakota Memories narrators reminisce about the colorful splendor of summer, as cultivated fields were lush and green. Rivers, streams, and wetlands also made this season quite unforgettable. Children found pleasure in nearby waterways, especially when the thermometer reached triple digits.[49]

The main reason children have fond memories of this warm season is the increased opportunity for outdoor activities. With the hot, humid days and parents and older siblings laboring in the fields, children were outside working, socializing, and playing. Many of the pastimes children took part in during spring carried over into the summer months. Plains children, for example, rarely tired of roaming the fields and pastures. During the summer months, children usually incorporated some time for playing in a nearby river or wetland, if possible. Only a few children actually knew how to swim. Most had fun just splashing around. According to some narrators, their fascination with waterways had more to do with the surrounding wildlife than the water itself.[50] Elaine Marie (Volk) Selensky, from McHenry County, North Dakota, shares a similar memory, recalling her childhood as simply wonderful. "Like everybody else," she recalls, she wandered the countryside barefoot. One of her favorite pastimes was going down to the nearby river to watch the beavers swim. Elaine says they

---

[47] Interviews with Nicolas F. Axtmann, Hilda Melita (Lindemann) Backfish, Polly Berger, Joseph John Black, Raymond Boechler, Wilfred Jacob Boechler, Edwin "Emil" Dewald, Anna (Sturn) Doll, Drefs, James Engelhart, Calvin K. Fercho, Marvin G. Fiechtner, ElRoy Fischer, Ruth (Dockter) Fischer, Emil Frank, Robert R. Gab, Martha M. (Brosy) Heyne, Nick Schall Jr., Rueben Richard Wolf, and Delmar D. Zimmerman.

[48] Interviews with Dewald, James Engelhart, Fercho, ElRoy Fischer, Ruth Fischer, Martha Heyne, Viola Konrad, Laturnus, Schaffer, Schall, Elaine Marie (Volk) Selensky, and Deloris Zimmerman.

[49] Interviews with Ruth Fischer, Martha Heyne, Jacob Johner, Konrad, Laturnus, Elaine Selensky, and Deloris Zimmerman.

[50] Interviews with Joseph Black, Emil Dewald, Drefs, James Engelhart, ElRoy Fischer, Ruth Fischer, Martha Heyne, Konrad, Laturnus, and Elaine Selensky.

had fun in the "good ole days!"[51] Summer, more than any other season, prompts such deep nostalgia.

For those who knew how to swim, summer was definitely a favorite season. Some towns had swimming pools, and some communities had large lakes or ponds. In south-central North Dakota, narrators have fond memories of Salt Lake.[52] According to *North Dakota: A Guide to the Northern Prairie State,* Salt Lake, which is officially called Lake George, had a "heavy impregnation with natural salts." This lake, as the guide also points out, had beautiful sandy beaches. It was because of these beaches that local German-Russians flocked to Salt Lake in the mid1900s.[53] It was a hotspot for family fun. Activities ranged from family picnics to sack races to swim competitions to dances. It was a social paradise, as narrators recall.[54] In fact, parents used a trip to Salt Lake as a reward (or bribe) for a good day's work. Ruth (Dockter) Fischer, from Streeter, North Dakota, recalls that after completing their chores, her parents would take the family to Salt Lake for a refreshing swim. "After we were all done with our chores, Mother and Dad would take us over to Salt Lake for a refreshing swim." These gatherings at Salt Lake provided a much-needed respite from the relentless summer heat and fostered a strong sense of community among German-Russian families. [55]

Parents also used Fourth of July celebrations as work incentives because children looked forward to this holiday with anticipation. It was second only to one holiday in their eyes: Christmas. Rural towns throughout the northern plains region held celebrations on the national holiday. The festivities typically involved communities organizing and hosting a parade, a picnic, games, and a dance. It was a time of socializing. It broke up the monotony of daily life. Since July is in the middle of summer, this celebration gave families the opportunity to enjoy the good weather.[56]

---

[51] Interview with Elaine Selensky.

[52] Interviews with ElRoy Fischer, Ruth Fischer, and Dockter.

[53] "North Dakota: A Guide to the Northern Prairie State," in *Workers of the Federal Writers Project of the Works Progress Administration for the State of North Dakota* (Fargo, ND: Knights Printing Company, 1938), https://www.gutenberg.org/files/46661/46661-h/46661-h.htm.

[54] Interviews with Dockter, ElRoy Fischer, and Ruth Fischer.

[55] Interview with Ruth Fischer.

[56] Interviews with Backfish, Doll, James Engelhart, Martha Heyne, Ida (Zimmerman) Koepplin, Philip Lang, Magdalena Miller, Eva (Helfrich) Mutschelknaus, Vernon Presler, Elda (Schulz) Rasch, Orion Arlyn Rudolph, Eugenia (Vetter) Schumacher, Simon Schwan, Gorden R. Werth, and Rueben Richard Wolf.

Playing ball was one of the most enjoyed pastimes of the summer months.[57] Games were played in both informal and organized settings. There were organized games at social events, such as the Independence Day and Canada Day celebrations, where communities gathered for friendly matches. Families and friends also played simple, casual games on farms or in towns. This pastime was not limited to summer—people started playing as soon as the snow melted in spring and continued until the first snowfall in late fall—so it was a multi-seasonal endeavor, spring, summer, and fall. Narrators look back on these recreational times with fond memories of family and friends. Rarely do they share memories about winning or losing; rather their memories are about togetherness. One of Viola (Eslinger) Konrad's fondest Fredonia, North Dakota, childhood memories is playing ball in the yard with her parents. Despite her mother often being pregnant, she and Viola's father always made time to join in, turning a simple game into a cherished family tradition.[58]

A few Dakota Memories narrators found pleasure in riding bicycles. Considering the summers were largely dry, conditions were perfect for this pastime—no snow or mud to slow one down. Not many children had this opportunity, as only a few families owned a bicycle before World War II. It was more common for children to ride horses for enjoyment than bicycles. According to the narrators who have memories of bicycle riding, it was an item shared by all children in the family. A bike was a hot commodity. Those who reminisced about riding bicycles and horses focused on the freedoms these activities provided, seeing summer as a season of exploration rather than limitation.

While Dakota Memories narrators share positive and happy memories of their summers on the northern plains, they also recall the challenges that came with the season. Farm labor was essential during the summer months, as pastures and fields remained frozen for nearly half the year. Several narrators describe the struggles of the drought years, the dust storms, the grasshoppers, and the prairie fires, which often threatened their livelihoods. Even as they cherish memories of summer's joys, they do not overlook its difficulties, reinforcing a narrative that values resilience and perseverance, and reflecting their desire to

[57] Interviews with Wilfred Boechler, Konrad, Peggy Sellinger, Mary Sellinger, Schaffer, Gilbert Ost, Mildred Wald, and Martha Heyne.

[58] Interview with Konrad.

maintain their ethnic group's story—one rooted in a shared history of endurance and survival rather than ease or abundance.[59]

Growing up in the first half of the twentieth century, the second and third generations experienced firsthand the weather extremes of the region. As narrators reminisce about their childhoods, they cannot help but acknowledge the hot summers. It was nearly impossible to stay cool, especially during the 1930s. There were days when the thermometer read more than 100°F. For a few narrators the relentless heat made summers on the plains nearly unbearable.[60] Mildred (Glatt) Wald, from Hague, North Dakota, says it best: "[Y]ou had to be tough" growing up on the northern plains. For Mildred, the summer heat was among the worst aspects. "You can put more clothes on" in the winter, but she points out in the summer "you can be naked and it's still hot."[61]

Children on the plains confronted severe heat waves and weather systems, like thunder and lightning, windstorms, and tornadoes. Dakota Memories narrators have vivid memories of thunderstorms, and a few even have memories of tornadoes. Tornadoes, however, were rare. Some narrators have recollections of winds gusting up to as much as fifty miles per hour. Others have memories of just the thunder rumbling and the lightning brightening the night sky. On countless occasions, summer storms damaged crops and homes, as well as terrified young children, as narrators attest.[62]

Despite the familiarity children had with thunderstorms and strong winds, most Dakota Memories narrators never experienced a truly horrific storm. While thunder, lightning, and wind were scary for children, these storms rarely caused them harm. Then again, occasionally a town or county newspaper ran a story, like "Lightning Kills Two in Emmons." According to the report, two German-Russians died near Strasburg, North Dakota. It states that Anton Mathern and his daughter, Anna, were struck by lightning on their farm. "The father was

---

[59] Interviews with Axtmann, Polly Berger, Joseph Black, Raymond Boechler, Drefs, Fercho, Marvin Fiechtner, Ruth Fischer, Hartmann, Jacob Helbling, Stanley Helbling, August Heyne, Martha Heyne, Jacob Johner, Lorraine (Meidinger) Kaseman, Kohlman, Laturnus, Schaffer, Schall, and Esther (Kauk) Schlecht.

[60] Interviews with Axtmann, Drefs, Ruth Fischer, Jacob Helbling, Jacob Johner, Schaffer, Schall, Schlecht, and Peggy Sellinger.

[61] Interview with Mildred Wald.

[62] Interviews with Axtmann, Polly Berger, Joseph Black, Raymond Boechler, Wilfred Boechler, Drefs, Hartmann, Jacob Helbling, Stanley Helbling, August Heyne, Martha Heyne, Jacob Johner, Lorraine Kaseman, Laturnus, Schall, Schmid, Peggy Sellinger, Mary Sellinger, and Delmar Zimmerman.

in the act of pitching hay," the story reads, "and the daughter with her pitchfork ready to receive same when the bolt from a small storm cloud struck." Two of Mathern's sons were raking nearby and witnessed this tragedy. Their father was forty-one, their sister fifteen. This tragedy was not an isolated event. While Dakota Memories narrators seldom experienced such a misfortune during their childhoods, stories similar to this one in the *Napoleon Homestead* in 1935 circulated throughout communities.[63]

Crop destruction was another obstacle the second and third generations faced, with strong winds being but one cause. Countless narrators have memories of insect and lagomorphs infestations. Grasshoppers, flies, and jackrabbits, to name a few, were quite a headache for farmers on the northern plains.[64] Stanley and Jacob Helbling from Morton County, North Dakota, experienced firsthand the devastation caused by grasshoppers. They both remember grasshoppers swarming their respective farms and devastating their fathers' crops. The grasshoppers gathered on the fence posts and ate the bark and nearly destroyed the fences. Stanley remembers them swarming the header box during lunch hour. Before he could head back to work, he had to clean it off. The chickens had a heyday eating the fallen grasshoppers. The infestations nearly destroyed the Helbling farms, as crop production declined dramatically during the 1930s. Stanley insists his family simply put up with it, because they had no other options.[65]

Grasshoppers were not the only annoying insects. For a few narrators, flies and mosquitoes were also a nuisance. Flies annoyed young girls the most, as the flies were especially bad in the kitchen (or summer kitchen—an outbuilding where families cooked, baked, and ate in order to keep the house cooler). Families had a number of remedies to keep the flies out, such as spraying insect repellent, hanging sticky tape from the ceiling, and placing linoleum strips in front of entryways. Regardless of the remedy, killing these irritating insects meant more work, as someone had to clean up the remains. Thus, women especially remember flies during their childhood because they created more hardships in their daily lives.[66]

---

[63] "Lightning kills two in Emmons," *Napoleon Homestead*, 26 July 1935.

[64] Interviews with Axtmann, Jacob Helbling, Stanley Helbling, Lorraine Kaseman, Schall, Schlecht, and Schmidt.

[65] Interviews with Stanley Helbling and Jacob Helbling.

[66] Interviews with Schlecht, Lorraine Kaseman, and Deloris Zimmerman.

The drought conditions of the mid 1900s also directly affected children's lives.[67] Marvin G. Fiechtner, from Wishek, North Dakota, experienced first-hand the horrors of the 1930s, also known as the Dirty Thirties and the Dust Bowl Era. Marvin recounts the dangers of dust storms in his oral history. In his words, a "choking dust" blew into town from time to time making it nearly impossible for children to play outside. There were occasions, according to Marvin, that these dust storms sent him running in search of a water faucet to rinse his mouth. The sand caked his mouth, making it nearly impossible to breathe, or so he felt. Narrators also describe the endless battle they faced keeping the house dusted and the farm equipment in working order. The fine dust particles infiltrated machinery, leading to increased wear and tear, frequent breakdowns, and costly repairs, thereby hindering essential farming operations.[68]

The increase in work and responsibilities continued from spring to summer. With warmer temperatures and dry land, parents needed children outside of the house more often. Narrators reminisce about their ever-present chores, like milking cows, rounding up horses, and feeding chickens. Considering that some children were responsible for the livestock year round, this chore was not seasonal. Some children only took care of the livestock in the summer and fall, when their fathers and older brothers were too busy laboring in the fields. Narrators—men and women alike—remember helping with watering, weeding, and spraying the gardens and crops during the summer months.[69]

Surviving summers on the northern plains was a character-building experience. Narrators reminisce or complain about a number of hardships including the droughts, heat, insect infestations, dust storms, and chores, but in spite of these hardships, most narrators still found time for enjoyments—taking excursions to Salt Lake, watching the beavers build dams, and playing a good old game of ball. For the most part, however, Dakota Memories narrators take care to emphasize their stories of obstacles, at the expense of their stories of opportunities.[70] By initially emphasizing their stories of hardships more than their sto-

---

[67] Interviews with Joseph Black, Raymond Boechler, Fercho, Marvin Fiechtner, Jacob Helbling, August Heyne, Jacob Johner, and Laturnus.

[68] Interview with Marvin Fiechtner.

[69] Interviews with Polly Berger, Joseph Black, Dewald, Drefs, Fercho, Jacob Helbling, Stanley Helbling, August Heyne, and Schaffer.

[70] Interviews with Axtmann, Polly Berger, Joseph Black, Drefs, Hartmann, Jacob Helbling, Stanley Helbling, August Heyne, Martha Heyne, Jacob Johner, Lorraine Kaseman, Kohlman, Konrad, Philip Lang, Laturnus, Schaffer, Schall, Schmidt, Elaine Selensky, and Peggy Sellinger.

ries of enjoyments, these German-Russian descendants are maintaining their ethnic group's established collective story—a tale of oppression and hardship. However, with deep interviewing, oral historians encourage narrators to reveal their individual memories. These memories reveal an altered narrative—a tale of opportunity, fun, and adventure.

Fall is a season that narrators from the Dakota Memories Oral History Project do not often discuss at length. When they do, their recollections typically revolve around harvest activities or the start of the school year. Similar to spring, fall on the northern plains is brief, lasting only a month or two, and is marked by increased workloads and a mild climate. Those who do share memories of this season acknowledge that, like other times of the year, fall presented both opportunities and challenges. During the fall season on the plains, people hunted and fished and took part in a number of other outdoor activities. Their opportunities for such activities were often limited as the crops needed harvesting, which was a labor-intensive and time-consuming process.[71]

Given the incredible variety of wildlife on the northern plains, people of the plains enjoyed the opportunities fall brought, such as hunting. Although narrators remember hunting being a year-round activity, fall was the best time of year for this pastime. The thrill of the hunt, the adventure of wandering the landscape, and the family togetherness are just a few aspects narrators remember fondly.[72] Some children even made extra spending money by guiding out-of-state hunters. Marvin Fiechtner, for instance, was quite proud of a childhood nickname he shared with his friend: The Great Hunters. These boys served as guides for a number of hunters searching for a new experience on the horizontal landscape of the northern plains. During his oral history session, Marvin also pointed out that occasionally these hunters left behind shotgun shells (and on one occasion a shotgun or rifle) to express their gratitude.[73]

As with hunting, trapping was a year-round sport at its height during fall season. Nearly all narrators have a memory or two of trapping animals, whether it was skunks, jackrabbits, or some other pesky critter. These animals

[71] Interviews with Alice Buck, LeRoy Buck, Dewald, Marvin Fiechtner, ElRoy Fischer, Ruth Fischer, Hartmann, Jacob Helbling, Stanley Helbling, Herman, August Heyne, Lorraine Kleingartner, Rupert Kleingartner, Lang, LeRoy H. Presler, Vernon Presler, Schlecht, Herbert Thurn, Delmar Zimmerman, Deloris Zimmerman.

[72] Alice Buck, Dewald, Marvin Fiechtner, ElRoy Fischer, Hartmann, Jacob Helbling, and Stanley Helbling.

[73] Interview with Marvin Fiechtner.

were quite destructive to farmers; as a result, people of the plains participated in mass hunts or roundups to eradicate them. These hunts usually took place sometime in late summer, early fall—before harvest. They involved anywhere from three to twenty-five people trapping and killing the animals.[74] Marvin L. Hartmann, from Delmont, South Dakota, participated in these mass hunts. When they were hunting for rabbits, Marvin says they usually split up to cover each section line, and then they converged toward the middle. The goal was to encircle the rabbits. Marvin recalls it was a sure-fire way to kill scores of rabbits, as well as foxes.[75]

Children, as Dakota Memories narrators reveal, found immense joy in simply roaming the vast expanses of the northern plains, foraging for wild fruits like chokecherries, wild grapes, and prickly pears, depending on the season. For many, these excursions were more than just berry-picking; they were adventures that fostered a deep connection to the land. Children also delighted in observing wildlife, from the playful antics of prairie dogs to the majestic flight of hawks. Domesticated animals, such as horses and cattle, provided endless fascination and companionship. These experiences not only offered entertainment but also instilled a profound appreciation for nature and the rhythms of rural life.[76]

For Dakota Memories narrators fall is synonymous with harvest season.[77] Philip Lang, from Tramping Lake, Saskatchewan, is one narrator who associates harvest with the opportunities of fall season. He found the harvest process simply beautiful and has fond memories of watching his father prepare the binder and other farm equipment. One of the best sights in the world, according to Philip, is a field full of stooks—bundles of grain arranged in a circular formation. He took pride in helping his father bring in a nice crop, so much so, that Philip continued in his father's footsteps, and in 2005 celebrated the one-hundredth year of his family's farmstead.[78]

Most Dakota Memories narrators did not find as much pleasure in harvesting as adults as they did when they were children. When asked to share

---

[74] Interviews with Hartmann, Herman, LeRoy Presler.

[75] Interview with Hartmann.

[76] Interviews with Alice Buck, LeRoy Buck, Dewald, Hartmann, Herman, Rupert Kleingartner, LeRoy Presler, Vernon Presler, Schlecht, and Deloris Zimmerman.

[77] Interviews with Drefs, Mathias "Math" A. Gross, Rupert Kleingartner, Kohlman interview; Philip Lang, and Schmidt.

[78] Interview with Philip Lang.

Image 2.5. Left to right, Ermaline Burlack, Alma Burlack, Eileen M. (Burlack) Kattlerling, Marion Burlack, and Arleen Burlack with a crop of corn, *circa* late 1930s, showing how farm work shaped children's identities and built pride in family self-sufficiency. *Photograph from the Eileen M. (Burlack) Kattlerling Collection, GRHC.*

their memories of fall season, narrators focus on the endless hours of labor and relentless heat. Harvesting required the labor of all family members. Crews of men and women worked in the fields to bring in the crop, while a few women were in the kitchen or the summer kitchen preparing meals and snacks. The summer kitchens, often a small building or shed adjacent to the main house, not only helped to avoid heating up the house; they also helped to reduce the risk of fire.

Usually, only the men and older boys operated the farm equipment while the women, girls, and younger boys assisted on the ground. Nonetheless, it was not uncommon for boys as young as ten or twelve years old to operate horse-drawn plows or gas-powered tractors. This process exhausted full-grown, healthy adults. Thus, harvesting was not a highlight of childhood for a number of Dakota Memories narrators.[79]

The labor associated with harvesting was not the only negative aspect of the fall seasons on the northern plains.[80] One of the most terrifying acts of nature (or humans) typically occurred in the late summer, early fall: prairie fires. Fortunately, only a few narrators had personal experience with prairie fires. Alice Ruth (Miller) Buck, born in Streeter, North Dakota, recalls a particularly harrowing childhood memory involving a prairie fire. One autumn day, when she was about seven or eight years old, she noticed a large cloud of smoke south of her school, moving northward toward her family's farmstead. Her teacher, recognizing the imminent danger, dismissed school early to ensure the children could return home safely. Upon arriving at her farm, Alice found her family and neighbors diligently plowing firebreaks around the house to prevent the fire's spread. Even years past the event, Alice's expression as she recounts this experience vividly conveys the fear she felt as a child. She specifically remembers her father's arm hair being singed while he was plowing the firebreaks. Fortunately, the fire burned out before reaching the Miller home, but the incident left a lasting impression on Alice.[81]

---

[79] Edwin "Emil" Dewald, interview; Jacob A. Helbling, interview; Stanley Helbling, interview; Herbert Thurn, interview.

[80] Interviews with Alice Buck, LeRoy Buck, Dewald, ElRoy Fischer, Hartmann, Stanley Helbling, August Heyne, and Herbert Thurn.

[81] Interview with Alice Buck.

Dakota Memories narrators clearly demonstrate by their memories a desire to maintain their collective identity—one shaped by survival and resilience. Nonetheless, their individual memories reveal a more complex storyline. While their ancestors endured political persecution, economic struggles, and forced migration, the second and third generations faced new challenges unique to life on the northern plains.[82] They adapted to extreme environmental conditions, navigated cultural assimilation, and balanced hard labor with moments of leisure. Unlike their immigrant ancestors, who relied on communal village life and traditional farming methods, these generations experienced a different kind of hardship—one that required self-sufficiency on isolated homesteads, learning new agricultural techniques, and integrating into a rapidly modernizing society. Their experiences—roaming the plains, hunting deer, trapping jackrabbits, and harvesting fields—reflect both continuity and adaptation. By sharing these positive experiences, as well as their negative experiences, these German-Russian descendants reinterpret their ethnic identity within the North American context, ensuring that their story remains one of perseverance and transformation.

Winter is the season most discussed by Dakota Memories narrators. They insist that winters on the northern plains were a true test of their ethnic group's will to survive and persevere. These months were long, cold, and at times dangerous. Countless narrators recall the frigid temperatures and severe blizzards. Nonetheless, narrators also acknowledge the beauty of winter on the plains—a winter wonderland, as some have referred to it. During these long five to six months, responsibilities were less intense, giving children more free time for activities. When the weather was manageable, German-Russian children took part in a few outdoor activities, such as sledding and ice-skating. During these months, however, children spent a good deal of the time indoors.[83]

During the harsh winters on the northern plains, German-Russian families often found themselves confined indoors due to subzero temperatures and biting winds. Living in cramped quarters with multiple children sharing rooms

---

[82] Interviews with Marvin Fiechtner, Alice Buck, LeRoy Buck, Dewald, ElRoy Fischer, Ruth Fischer, Hartmann, Jacob Helbling, Stanley Helbling, Herman, August Heyne, Lorraine Kleingartner, Rupert Kleingartner, Philip Lang, LeRoy Presler, Vernon Presler, Schlecht; Herbert Thurn, Delmar Zimmerman, and Deloris Zimmerman.

[83] Interviews with Clarence Bauman, Polly Berger, Joseph Black, Raymond Boechler, Wilfred Boechler, Alice Buck, LeRoy Buck, Dewald, Clara (Ell) Ferner, Fercho, ElRoy Fischer, Magdalena (Staudinger) Hauck, Wilma (Beck) Haupt, Jacob Helbling, Herman, August Heyne, Louis Helfrich, Jahner, Jacob Johner, and Ruth Rosa Johnson.

and even beds, they relied on simple indoor games to pass the time. One such game was "Drop the Handkerchief," where children formed a circle, and one child walked around the outside, stealthily dropping a handkerchief behind another. The chosen child would then pick it up and chase the first around the circle, aiming to tag them before they could occupy the vacant spot. Playing cards were also popular; however, some religious beliefs prohibited their use due to associations with gambling and moral concerns. For instance, certain religious denominations viewed card playing as a sinful activity that could lead to immoral behavior. Despite these indoor diversions, children eagerly awaited opportunities to venture outside and play in the snow whenever the weather permitted.[84]

Growing up on the northern plains, the second- and third-generation German-Russians knew the proper attire needed to enjoy the winter season. The wardrobes were simple yet warm and practical. Narrators recall gender distinctions in their clothing—girls typically wore dresses, often layered over warm undergarments such as long woolen stockings or leggings to provide insulation against the cold. They also wore knitted sweaters and shawls for added warmth. Boys, on the other hand, wore jeans or bib overalls paired with sweaters and jackets. Boys and girls commonly wore knitted hats and mittens, often handmade by family members, to protect against the frigid temperatures. Given the financial situation of plains people in the 1920s and 1930s, as well as the average German-Russian family size of ten to twelve children, narrators remember wearing mostly hand-me-downs. These clothes must have worked, because countless narrators have fond memories of a winter paradise on the northern plains.[85]

Winter ice and snow provided opportunities for different pastimes, like sledding, skiing, and skating often during the noon-hour or recess at school and at home in the evenings after supper and chores. Children could spend endless hours sledding and skiing down small hills or snow-covered haystacks on their family farms. If parents and older siblings would have allowed it, children would have stayed out all night. Skating on nearby lakes was also quite enjoyable. With the extended winter season and numerous wetlands, the environment provided plains children with natural ice-skating rinks. Some children never had store-bought ice skates, but homemade ones worked well enough. According to Da-

[84] Interviews with Joseph Black, Philip Lang, and Mary (Mitzel) Weber.

[85] Interviews with Joseph Black, Drefs, Jacob Helbling, Ruth Rosa Johnson, Rosa Laturnus, Julius Miller, Peggy Sellinger.

kota Memories narrators, they never really felt the cold when they were playing. They were always bundled-up and having too much fun to notice.[86]

Hockey was a favorite pastime in Canada. The frozen wetlands and sloughs provided the perfect and sometimes dangerous outdoor hockey rink. A handful of narrators were fortunate enough to have store-bought hockey equipment, including skates, sticks, and pucks. However, homemade equipment also worked: sticks were turned into hockey sticks, rocks into pucks, and shoes into skates. Plains children found imaginative ways to participate in winter pastimes. Even the labor associated with clearing the ponds of fresh snow rarely deterred them. Narrators remember spending hours shoveling off lakes, ponds, or sloughs in order to play. Yet, for all the joy winter brought, it also came with immense challenges—harsh temperatures, heavy snowfall, and blizzards that could isolate entire communities for days.[87]

While a few narrators have fond memories of winters on the northern plains, most prefer to reminisce about the hardships they overcame during these frigid, unpredictable months. They were cold. There was lots of snow. Blizzards were quite common, and sometimes lasted for two or three days. Some narrators recount the frustration they experienced attempting to keep the house warm during the winter months. It was common for the kitchen to be the only warm room in the house. As they share their memories, a few narrators even pause, thinking that maybe they are exaggerating the hardships. Then they quickly recant, insisting it is not an exaggeration, because their memories are so vivid.[88]

Blizzards and excessive snow accumulation made daily chores increasingly difficult. The snowbanks were at times as tall as the houses, barns, and outbuildings, making it difficult to navigate the farmstead. Animals needed feeding and most rural homes lacked running water and indoor bathrooms, conveniences that were not available until the 1950s and 1960s. One attempt at safety, used with some frequency, was to tie ropes connecting the outhouse to the house and to the barn to help guide family members and friends across the

---

[86] Interviews with Joseph Black, Raymond Boechler, Alice Buck, Fercho, Jacob Helbling, Herman, Jahner, Jacob Johner, Lorraine Kaseman, Konrad, Christina Rose Marie (Bast) Krismer, and Deloris Zimmerman.

[87] Interviews with Raymond Boechler, Wilfred Boechler, Jahner, and Jacob Johner.

[88] Interviews with Raymond Boechler, Wilfred Boechler, LeRoy Buck, Dewald, Marvin Fiechtner, ElRoy Fischer, Ruth Fischer, August Heyne, Martha Heyne, Jacob Johner, Johnson, Lorraine Kleingartner, Rupert Kleingartner, Kohlman, Peter A. Kraft, Rosa Laturnus, Christina (Alberle) Long, Julius Miller, Schall, and Gust Wanner.

farmstead in blizzardy weather. Occasionally, families built steps or tunnels in the snow. The creativity and ingenuity of parents made it possible for families to survive the long cold winters on the northern plains.[89]

Getting to the barn and outbuildings was never as bad as getting to school, according to Dakota Memories narrators. During the winter months, the one- to three-mile walk children faced became unbearable. When winter came, families used alternative transportation to get children to and from school.[90] Born in Sheridan County, North Dakota, Ruth Rosa (Korus) Johnson vividly recalls using a *budke*—an enclosed sled pulled by horses for winter transportation. The term *budke* was commonly used among German-Russian families to describe this type of bus, though it may not have a direct translation in standard German. The budke shown in Image 2.4 was used by the Korus family; it was rounded with only one window in the front, through which a driver could navigate and handle the horses from inside the shelter. The budke also had a single door in the rear for people to enter and exit. Ruth remembers her father putting a foot warmer inside the budke. The warmer was a metal box, full of hot charcoal. Ruth acknowledges that in hindsight, this may have been dangerous, but they never had any problems. Of course, as Ruth notes, the Korus children also dressed properly and snuggled into blankets for their rides to and from school. This resilience in the face of harsh conditions reflects the enduring spirit of their German-Russian heritage, where enduring and overcoming adversity was integral to their collective identity. Reinforcing this ancestral narrative of perseverance, Ruth says it was cold, but they survived.[91]

While Dakota Memories narrators might complain about the long, cold winter months, most never experienced a weather-caused tragedy. Still, newspapers documented that there were tragic instances such as "Lehr, Kulm Youths Severely Frozen" and "Emmons Man is Blizzard Victim." These incidents serve as stark reminders of the unpredictable and often unforgiving nature of northern plains winters, where a routine trip could quickly turn life-threatening. The first story mentioned here, published on February 4, 1938, claims that "eight youths, four Kulm girls and four Lehr boys," all from German-Russian families,

---

[89] Interviews with Raymond Boechler, ElRoy Fischer, Jacob Johner, Krismer, Long, Schaffer, Schlecht, Wanner, and Delmar Zimmerman.

[90] Interviews with Wilfred Boechler, LeRoy Buck, Dewald, James Engelhart, ElRoy Fischer, Ruth Fischer, Jacob Helbling, Johnson, Lorraine Kleingartner, Rupert Kleingartner, Konrad, Peter Kraft, Philip Lang, Rosa Laturnus interview, Long, Schaffer, Schumacher, and Wald.

[91] Interview with Johnson.

Image 2.6. The Koruses' budkie, also known as the family school bus, *circa* 1930s, demonstrating the ingenuity and determination of German-Russian families to ensure their children received an education despite harsh prairie conditions. *Photograph from the Ruth Rosa (Korus) Johnson Collection, GRHC.*

"were seriously frozen about the limbs and face Saturday night when their car stalled in a snowbank four miles east of Wishek." The report continues with the grim story: "[T]he girls' limbs were frozen up to their hips. Ears, noses and hands were frostbitten." Stranded in the elements with no immediate help, the group endured severe frostbite, demonstrating how quickly conditions could escalate into dangerous situations. The second story, which ran on February 17, 1939, was accompanied by a descriptive subtitle: "Adam Vetter, Farmer near Hague, Disregarded Advice to Stay in Town." The story reads: "Adam Vetter, 45, farmer near Hague, lost his life in last week's storm. His frozen body was found in the snow after he had attempted to walk to his farm from the town of Hague."[92] Unlike the stranded youths, who survived despite their injuries, Adam Vetter's decision to face the storm alone proved fatal, reinforcing the ev-

[92] "Lehr, Kulm Youths Severely Frozen," *Napoleon Homestead*, 4 February 1938; "Emmons Man is Blizzard Victim," *Napoleon Homestead*, 17 February 1939.

er-present risks posed by blizzards to even the most experienced plains residents. These stories, while rare, are examples of the unpredictable nature of the northern plains environment.

Rupert Kleingartner, born in Gackle, North Dakota, is one of only a few narrators who experienced such a tragedy because of a North Dakota winter. When he was twelve years old, his mother died in childbirth during a blizzard. She was carrying twins and went into premature delivery. The weather was so bad that the family had no way of getting a midwife, much less a doctor, to the farmstead. Rupert recalls the hired man hitching up the horses and fetching the neighbor lady to assist in the birthing process. Rupert's mother, Alga (Schlenker), gave birth to two boys, each weighing less than five pounds. The boys were so small that when they were born his father put them in shoeboxes and used the kitchen stove to keep them warm. For a short time, they thought one of the babies might have been stillborn, but he eventually started to cry. Alga started hemorrhaging shortly after labor; she died eight days later. The loss of his mother was truly an earth-shattering experience for Rupert. Her death was not just a personal tragedy—it was a stark illustration of how winter could take lives in ways beyond frostbite or exposure. The inability to reach medical assistance, even days after a storm, was one of the many dangers German-Russian families faced in the unforgiving winters of the northern plains.[93]

Some say surviving winter on the northern plains builds character, a statement to which second- and third-generation Germans from Russia attest. Surviving subzero temperatures without the modern conveniences of electricity and central heat was hard. In the early to mid-1900s, rural plains people coped with a number of hardships, including severe winter weather. As German-Russians tell it, winters were just one of many obstacles they faced and overcame, with grace. Furthermore, Dakota Memories narrators insist they are better for having experienced these hardships.[94]

Whether reminiscing about spring, summer, fall, or winter, Dakota Memories narrators share stories of both hardships and opportunities. Life on the plains as children, in other words, was both frustrating and rewarding. They had to adapt and cope with the unpredictable and severe weather on the plains, and they had to deal with unpleasant insects and animals. Yet, there were so

---

[93] Interview with Rupert Kleingartner.

[94] Interviews with Polly Berger, Joseph Black, Raymond Boechler, Wilfred Boechler, Alice Buck, LeRoy Buck, Dewald, Drefs, James Engelhart, Fercho, Marvin Fiechtner, Jacob Johner, Johnson, Lorraine Kaseman, Krismer, Philip Lang, Long, Julius Miller, Weber.

many peaceful and worthwhile aspects of the environment, too, like growing up in a wide-open area with hills to roam and fields to wander. A number of their pastimes also involved interacting with the environment in a variety of ways, such as trapping, hunting, sledding, ice-skating, and swimming. Through these means and experiences, they—as Alice Buck concluded— found "beauty on the plains."[95]

Environmental experiences were not unique to Germans from Russia; children from various backgrounds, such as Wallace Stegner, Lois Phillips Hudson, and Dorothy Hubbard Schwieder, had similar interactions with the northern plains landscape. They, too, recalled the desolate terrain, choking dust storms, swarms of grasshoppers, and extreme temperatures and winds. Like the German-Russian children, they engaged in exploring the countryside, swimming in rivers and ponds, trapping gophers, and picking flowers. Highlighting these shared experiences emphasizes that, regardless of ethnicity, children growing up on the northern plains between 1900 and 1950 faced both positive and negative environmental interactions. This commonality underscores the universal challenges and joys of rural life during that era, fostering a deeper understanding of the region's cultural and environmental history.

Nonetheless, this analysis of German-Russian descendants encountering the northern plains environment complements, as well as expands, the existing historiography on children and the environment. It reaffirms the conclusions of Elliott West that children are historical actors interacting with their environment. According to West, the environment "was more than a background of the children's story; it was also an active force that helped shape their values and attitudes, their views of the future, and their understanding of themselves."[96] Because German-Russian descendants consider themselves people of the plains, they provide vital opportunities for historians to examine the interplay between German-Russian children and nature. The oral histories collected through the Dakota Memories project make such a unique study possible.

This research is also the first in-depth historical analysis of children on the northern plains in the twentieth century. While scholars like Elliott West exam-

[95] Interview with Alice Buck.

[96] West, *Growing Up with the Country*, 1.

ine children in the West during the settlement era, and others such as Hampsten and Holt have focused on nineteenth-century plains children, *Echoes of the Old Country* adds a new perspective by concentrating on German-Russian children's experiences in the northern plains during the twentieth century.[97] This addition to the existing historiography reveals both similarities and differences in childhood memories and experiences across time and space in the American West and specifically on the northern plains.

With collective memory rooted in inherited tales of hardships and sadness, Dakota Memories narrators altered their ethnic group's narrative to accommodate their new environmental experiences, thus creating their own new story. As children, they viewed the northern plains as their *heimat* (homeland). Thus, the German-Russian story is no longer solely about persecutions, hardships, isolation, and sadness. Their story is now one of opportunities and adventure, as they now incorporate both the tranquil and the terrifying.

Nowhere is this balance more evident than in the memories of individuals like Rosa (Schneider) Laturnus, who was born in Scott, Saskatchewan. She recalls both the beauty and the brutality of winter on the plains. In sharing her childhood memories of overcoming environmental hardships, she tells of the harsh realities of winters in the 1930s from firsthand experience. Her father constantly worried about her safety while she traveled to and from school, ever fearful of the dangers posed by frigid temperatures and unpredictable storms. So much so that he sometimes insisted on driving her to school with the horse and sled. The cold, she assured our Dakota Memories team, was nearly unbearable. She and her father wore overcoats, long stockings, mittens, and overshoes, but it was nearly impossible to stay warm. Rosa also has memories of blizzards lasting for two to three days, preventing them from feeding the cattle. In those moments, all they could do was wait. "You just had to let it blow over," she recalled, underscoring the patience and endurance required to survive winter on the northern plains.[98]

[97] For more information, see West, *Growing up with the Country*; Hampsten, *Settlers' Children*; Holt, *Children of the Western Plains;* Riney-Kehrberg, *Childhood on the Farm.*

[98] Interview with Rosa (Schneider) Laturnus.

CHAPTER 3

More Than a Photo: Family Relationships

"The thing I remember about [my grandmother]," Deloris (Boschee) Zimmerman, from Lehr, North Dakota, says is that "when you hugged her you couldn't even get your arms around her, but she was so soft, and she was always pleasant and so nice." Deloris spent much time with her paternal grandmother, Katherine (Rudolf) Boschee. She remembers her as a short, round woman. Her paternal grandfather, Jacob Boschee, she does not remember quite as well but recalls him as a quiet man who spent most of his time laboring in the fields. Nonetheless, her most treasured memories are spending the holidays with them. "At Christmastime," she recalls, they "always had candy for every one of the grandchildren, and they weren't wrapped in anything special. It was just a little brown bag of candy." [1]

Most Dakota Memories narrators reminisce a great deal about their family relationships—grandparent-grandchild relationships, parent-child relationships, and sibling-sibling relationships. They share memories of their grandmothers baking them delicious dough dishes, of their fathers making them labor endlessly in the fields, and of their siblings coaxing them into mischief. Through these relationships, German-Russians learned about their ethnic group's identity, and they also reshaped it into a new identity, blending their past with their present. Within the context of family relationships, these German-Russian descendants forged a new identity rooted in their ethnic heritage and history but serviceable to new, American- and Canadian-born generations. The story is now much more complex as it blends the traditional narrative of hard times and persecutions with memories of love and warmth.

[1] Interview with Deloris (Boschee) Zimmerman.

Narrators of the Dakota Memories insist that family relationships played a vital role in the construction of their individual identity, as well as their collective identity. Most European immigrants who settled the Great Plains region came to America and Canada as part of a family unit. Once in their new *heimat* (homeland), as historian Frederick C. Luebke acknowledges, these immigrant families typically retained their cohesiveness. In fact, families settled in tight-knit communities with other families from similar backgrounds, ethnicities, and religions.[2] German-Russians in particular settled in close proximity to their extended families—great-grandparents, grandparents, aunts, uncles, and cousins.

Family, as childhood and family historians note, is a social construct that varies in place and time. As a social construct, family exists because people imagine it. In their landmark work, *Family Fortunes*, social historians Leonore Davidoff and Catherine Hall describe the basic characteristics of modern European families. According to their findings, modern Europeans consider their parents and young children as part of their immediate family, and their grandparents, aunts, uncles, cousins, and adult siblings as part of their extended family.[3] Although most Germans from Russia were no longer living in Europe by the turn of the twentieth century (and they postdate Davidoff and Hall's research by more than one hundred fifty years), they view the construction of family in a similar manner. For Germans from Russia, family largely denoted a biological kinship. That is, it typically did not include social kinships.[4]

Family relationships played a vital role in child development and identity formation, involving both intergenerational and intragenerational interactions. Intergenerational relationships were those connections between children and older family members outside their generation, including great-grandparents, grandparents, parents, aunts, and uncles. These relationships shaped children's identities primarily through the transmission of cultural knowledge, traditions, and values. Anthropologists and historians have long noted that elders—particularly grandparents—acted as custodians of ethnic heritage by sharing stories, language, religious beliefs, and customs, thereby instilling a strong sense of

[2] Frederick C. Luebke, "Ethnic Group Settlements on the Great Plains," *Western Historical Quarterly* 8 (1977): 405.

[3] Leonore Davidoff and Catherine Hall, *Family Fortunes: Men and Women of the English Middle Class 1780–1850*, revised ed. (London: Routledge, 2002), 31.

[4] Dee Garceau-Hagen, "C.W.W.H. Roundtable Gender, Region, and Culture: New Voice and Directions," at *Western History Association*, Denver, CO (October 7, 2009).

belonging and cultural continuity in the younger generations. Parents further reinforced this cultural identify through daily interactions, providing practical discipline, work ethics, and moral guidance, all of which rooted firmly within their German-Russian cultural frameworks. Intragenerational relationships, on the other hand, were connections among children of the same generation, such as siblings and cousins. These relationships contributed significantly to children's development by providing spaces for mutual exploration, shared experiences, and collective memory-making, enabling them to reconstruct and adapt their ethnic identities within their contemporary environment.[5]

Family historians and specialists have often overlooked intergenerational relationships, especially the grandparent-grandchild relationship. Even less attention has been given to the relatively rare interactions between great-grandparents and great-grandchildren. Historically, prior to the twentieth century, this intergenerational dynamic was characterized by distant authority, meaning that older generations typically related to younger generations through formality, limited emotional intimacy, and clearly defined hierarchical roles. Elders were respected bearers of tradition and authority rather than close companions or confidants. However, with the evolving concept of childhood throughout the twentieth century—which emphasized emotional closeness, nurturing, and child-centered relationships—the grandparent-grandchild bond transformed significantly. Relationships once defined by formality and distance became increasingly characterized by warmth, friendliness, and mutual affection. This shift allowed grandparents to take on a reflective and supportive role, actively transmitting cultural knowledge, family stories, and ethnic traditions, thus deeply influencing their grandchildren's identity formation.[6]

Family historians and specialists have also underexplored the nuances of parent-child relationships, particularly how practical and authoritative parenting shaped childhood experiences. The parent-child relationship was practical in nature, providing children with education and life experience. Occasionally, children experienced a similar relationship with members of their extended family. This relationship depends on the authoritarian role of adults and the obedience of children. In a work about eighteenth- and nineteenth-century English families, historians Davidoff and Hall note that parental duties consisted

---

[5] Sonya Salamon, *Prairie Patrimony: Family, Farming, and Community in the Midwest* (Chapel Hill: University of North Carolina Press, 1992), 50–56.

[6] Crystal C. Ramirez Barranti, "The Grandparent/Grandchild Relationship: Family Resource in an Era of Voluntary Bonds," *Family Relations* 34 (1985): 346.

of "enforcing discipline and instruction in serious subjects."[7] This remained so in the twentieth century. Parents were preoccupied with their daily responsibilities, and as a result, were unaware of, or indifferent to, the changing concept of childhood. The primary purpose of the parent-child relationship remained to help children maintain and develop core values through discipline and hands-on training.[8]

Family and childhood historians have also largely overlooked the importance of children's intragenerational relationships, such as those between siblings and cousins. These relationships were distinct from interactions with adults, which tended to reinforce older traditions and norms. Yet, through interactions with peers, children found space to explore their environment, formulate their own perspectives, and enjoy their childhood on their own terms. By sharing discoveries and adventures, they did not merely replicate the stories of their elders but created new ones, reconstructing their identities through the shared experiences of their generation.[9]

This resilient, reconstructed narrative becomes quite evident in the Dakota Memories oral histories. Narrators spend plenty of time reminiscing about family life. There are at least two possible explanations for this focus on family history. First, preserving family history is one of the main objectives of the Dakota Memories project (as a result, the oral history teams have prepared several questions regarding family relationships). Second, the desire of German-Russians to preserve their traditional story stems from their familial ties (and therefore, their own plan generates an emphasis on family relationships).[10] Because of these overlapping agendas, the stories about family history and ethnic identity gleaned by the Dakota Memories teams are quite striking and provide insight into the new German-Russian story.

Although the relationship between great-grandparents and great-grandchildren was rare, primarily due to shorter lifespans and generational gaps, when it existed, it provided children with a unique and often fleeting connection to the distant past. These brief interactions offered children a reflective glimpse into their deeper ancestral roots, as great-grandparents shared firsthand mem-

---

[7] Davidoff and Hall, *Family Fortunes*, 31.

[8] More information, see Colin Heywood, *A History of Childhood: Children and Childhood in the West from Medieval to Modern Times* (London: Polity Press, 2001).

[9] For more information, see Annette Atkins, *We Grew up Together: Brothers and Sisters in Nineteenth Century America* (Urbana: University of Illinois Press, 2001).

[10] For more information on the Dakota Memories' objectives, visit: https://digitalhorizonsonline.org/digital/collection/p16921coll13.

ories and traditions that even grandparents might not remember clearly. Thus, despite their brevity, these interactions served as powerful moments of cultural continuity and identity formation for children, linking them directly to their family's distant past. Through these relationships, German-Russian children learned the basics about their ethnic heritage and the cycle of life.[11] Nonetheless, only a few narrators experienced this relationship.[12]

Born in Bismarck, North Dakota, in 1925, Calvin K. Fercho was one of few Dakota Memories narrators fortunate enough to have a brief relationship with a great-grandparent. Calvin describes his great-grandmother as a "prim" and "proper" German lady. He remembers her wearing long black dresses and speaking the German language. He insists she never spoke a word of English. According to his childhood memories, Calvin's great-grandmother was a stereotypical German-Russian woman.[13]

Selma (Job) Lapp, from McPherson County, South Dakota, didn't know her great-grandparents, but she makes assumptions about them based on stories passed down through her family. According to the family tradition her great-grandmother Job died from homesickness in 1902, twenty-five years before Selma was born. Selma's great-grandparents had emigrated from Russia to America after their son Ludwig encouraged them to join him on the northern plains. This migration was indirectly forced, prompted by Ludwig's changing political and military circumstances in Russia, including the removal of privileges and autonomy originally promised by Tsarina Catherine the Great. Selma's great-grandparents settled near their son in Eureka but struggled profoundly with adjusting to life in America, eventually succumbing to what the family described as homesickness. Unable to provide further detail, Selma acknowledges that she regrets not asking her father more questions about her family's past. Nonetheless, the poignancy of this narrative highlights how migration could become emotionally burdensome and even traumatic, reinforcing within the family memory an understanding of their ethnic history as rooted in displacement and hardship.[14] While not forced migration in the strictest sense, the pressures that led Germans from Russia to leave their homeland—loss of autonomy, cultural repression, and deteriorating living conditions under the Russian government—created circumstances akin to forced migration. Thus,

---

[11] Barranti, "The Grandparent/Grandchild Relationship," 348.

[12] Interviews with Fercho, Selma (Job) Lapp, Simon Schwan, and Shirley Ann (Weigum) Stern.

[13] Interview with Fercho.

[14] Interview with Lapp.

the family story of Great-Grandmother Job serves as a symbolic representation of the broader historical narrative of displacement and emotional hardship experienced by Germans from Russia.

One of the most significant twentieth-century German-Russian intergenerational relationships was the grandparent-grandchild relationship. Since this ethnic group largely settled in homogenous communities with family members and friends, German-Russian descendants knew their grandparents well. This relationship provided the younger generation with a connection to family history, cultural heritage, and collective ethnic identity. In addition, grandparents were role models or mentors for their grandchildren, teaching them gender roles and social expectations. Children also learned about the life cycle through this relationship, as they inevitably lost grandparents during their childhoods.[15] In other words, this connection was influential for Germans from Russia growing up on the northern plains, as it provided them with something that parents were unable to supply: a direct connection to their ethnic heritage and ancestral roots.[16]

Countless Germans from Russia preserved their family history as a result of the grandparent-grandchild relationship.[17] Dakota Memories narrator Leona (Miller) Klundt, who was born in Gackle, North Dakota, remembers learning about her family history from her maternal grandparents. According to Leona, her grandparents, Johann and Christina (Gienger) Kammerer, were influential throughout her childhood. Leona and her younger sister, Alice Ruth (Miller) Buck, were always excited to visit their grandparents because they usually learned so much about their family history during their visits. For instance, Leona remembers listening to stories, in German, about her grandfather Kammerer's life in Russia. Her grandfather shared his memories not only about fleeing Russia with nothing but clothes and money, but also about losing his first wife, the mother of his three eldest daughters, and then marrying Leona's grandmother, Christina. Leona and Alice's grandparents provided her with a direct connection to her ancestral roots through their storytelling.[18]

---

[15] Barranti, "The Grandparent/Grandchild Relationship," 347–48.

[16] Interviews with Margaret Bitz, Adam Gordon Boschee, Alice Ruth (Miller) Buck, Esther (Haerter) Drefs, Raymond Erbele, Clara (Ell) Ferner, Alvin E. Graf, Marvin L. Hartmann, Jacob A. Helbling, Stanley Helbling, Peter Herauf, August Heyne, Geraldine (Dukart) Keen, Reinhold R. Klein, Leona (Miller) Klundt, Christina (Aberle) Long, LeRoy H. Presler, Valeria (Wagner) Sehn, James Vetter, and William "Bill" Wald.

[17] Interviews with Bitz, Graf, Jacob Helbling, Stanley Helbling, Klundt, Heyne, Keen, James Vetter, Wendelyn Vetter.

[18] Interviews with Klundt and Buck.

Image 3.1. Four generations of Schlecht women; left to right, Florence, Magdelina, Grandmother Becker, and Esther (Kauk) Schlecht, *circa* 1950, embodying the continuity, strength, and vital role of women in preserving German-Russian heritage. *Photograph from the Esther (Kauk) Schlecht Collection, GRHC.*

The grandparent-grandchild relationship inspired second- and third-generation Germans from Russia to maintain their ethnic identity.[19] James Vetter, hailing from Logan County, North Dakota, participated in Dakota Memories in order to share his family's immigration stories that his grandmother handed down to him. Jim recalls his grandmother, Magdelena (Unser) Vetter, visited often during his childhood, sometimes staying for several weeks at a time. During these visits, his grandmother shared stories about her family's life in Russia and migration to the United States. She reminisced about the happy times she had growing up in a *dorf* (a Russian village) and helping with communal farming.

---

[19] Interviews with Jacob Helbling, Stanley Helbling, Heyne, Keen, James Vetter, and Wendelyn Vetter.

Jim is certain that if the Russian army had not conscripted his grandfather, his grandparents would have remained in Russia.[20]

Jim's older brother, Wendelyn, also recalls stories passed down from older generations, which provided context for understanding their ethnic identity. Specifically, he remembers family narratives about their ancestors' struggles, which reinforced a sense of belonging to a resilient and determined community. One such story involved their Great-Aunt Mary Ann, sister of Jim and Wendelyn's grandfather, John, who faced challenges during immigration. In family accounts, Mary Ann was blind, and her disability initially caused immigration officers to question whether to allow the Vetter family into the United States during their entry process. Eventually—though it is unclear how—officers permitted them to continue their journey to Strasburg, North Dakota.[21] Stories like these taught Jim and Wendelyn not only about the historical trials their family overcame but also instilled core values central to German-Russian ethnic identity: resilience in adversity, persistence in the face of challenges, and family solidarity during hardship. By hearing about and empathizing with their ancestors' struggles, Jim and Wendelyn developed a deeper emotional connection to their German-Russian heritage, viewing themselves as part of an enduring tradition of strength and adaptation in the face of adversity.[22]

German-Russian descendants learned about their family history through their relationship with their grandparents, who actively shared and preserved ethnic traditions, stories, and values.[23] These interactions reinforced emotional bonds between generations while embedding children in a rich cultural heritage. For example, Christina (Aberle) Long, from Emmons County, North Dakota, vividly recalls one Christmas when her grandparents—Anthony and Francis (Vetter) Aberle—introduced her and her siblings to traditional German-Russian characters: the Belznickel, who frightened children into good behavior, and the Krist Kindel (Christ Child), who brought joy and reassurance. The Belznickel, a figure meant to instill obedience and respect, symbolized the disciplinary aspects of German-Russian upbringing, whereas the Krist Kindel embodied generosity and comfort, nurturing a sense of security and community. During this particular celebration, the Aberle children lined up anxiously

---

[20] Interview with James Vetter.

[21] Interview with Wendelyn Vetter.

[22] Interviews with James Vetter and Wendelyn Vetter.

[23] Interviews with Adam Boschee, Raymond Erbele, Clara Ferner, and Long.

on a bench, awaiting the arrival of these traditional figures. Christina believed the Belznickel would kidnap naughty children, so she was particularly relieved when the Krist Kindel, accompanied by a donkey, arrived bearing gifts. She remembers receiving a special cookie, a rare treat meant to reward good behavior. However, when the donkey approached her and showed interest in her cookie, Christina offered it to the animal instead of eating it herself. Her act of giving up the treasured cookie, driven by the powerful emotional contrast of fear and relief brought by the traditional Christmas characters, highlights how deeply these familial traditions influenced her personal values, demonstrating obedience, generosity, and empathy—core elements of her ethnic identity fostered through her relationship with her grandparents.[24]

German-Russian descendants also learned about distinctly ethnic traditions through holiday celebrations with their grandparents, experiences that also reinforced broader lessons about discipline, respect, and obedience valued by their families. Adam Gordon Boschee, born in Kulm and raised in Lehr, North Dakota remembers visiting his grandmother Margaret Kautz's home during holidays. Adam recalls that on one particular holiday, although he is not sure which one, his family rounded up all the young children and put them in a small room. Soon after, a short, round woman dressed in a white sheet and carrying a whip entered the room. Adam vividly remembers that she knew everything you had done wrong; she could "point out to you some things that you had recently did to make your mom mad. You wondered how." The woman, playing a disciplinary figure from traditional German-Russian holiday lore, would then lightly switch the children. Adam recalls "It wasn't real bad, but it would sting." This tradition was not merely about punishment; it symbolized and reinforced the high expectations parents and grandparents held for behavior in a community where life was often difficult, requiring disciplined and responsible children. Through such cultural rituals, older generations taught younger generations to appreciate the seriousness of family obligations and the consequences of their actions—critical lessons for survival in the demanding environment of the northern plains.[25]

This significant intergenerational relationship also taught children about their community's social expectations.[26] For instance, German-Russians ex-

---

[24] Interview with Long.

[25] Interview with Adam Boschee.

[26] Interviews with August Heyne, Bill Wald, and Deloris Zimmerman.

pected their young to respect their elders. German-Russian grandparents may have been more receptive to the twentieth-century construct of childhood (i.e., Viviana Zelizer's "emotionally priceless" child) yet they still valued traditional practices.[27] Old habits die hard. Deloris (Boschee) Zimmerman (Adam's sister), for example, remembers her grandparents demanding that children always show them the utmost respect, especially when conversing with them. When Deloris was a child, her family primarily spoke German. Thus, grandchildren were never to use the word *du*—the informal version of you—when conversing with their grandparents; they were supposed to use the formal version: *Ihr*. According to Deloris, if they ever forgot and used du instead of Ihr, there was serious trouble, because that was disrespectful.[28] This group's dedication to tradition combined with its willingness to change helps explain the resilience of the German-Russian ethnic identity. Germans from Russia accepted change, yet it was on their terms.

Another social expectation in German-Russian communities, one commonly inculcated by grandparents, was religious devotion.[29] William "Bill" Wald, from Venturia, North Dakota, witnessed firsthand the religious devotion of his grandparents. He insists that his maternal grandparents, Jacob and Katherina (Horner) Moch, were true role models. "They never missed a Sunday at church," Bill exclaims. Growing up during the 1930s, he remembers hearing about his grandparents traveling by horseback for five hours in blizzard conditions to attend a one-hour church service. His grandfather, as the family lore maintains, placed rocks, which he had heated in the oven, in the center of the buggy for his wife and children to stay warm. Bill insists it was almost impossible to deter his grandparents from attending church, and he credits them for teaching him to be a devout Catholic.[30]

This significant relationship also taught second- and third-generation German-Russians about the cycle of life. Inevitably, many German-Russian children experienced the loss of a grandparent during their childhood. This loss was frightening and confusing for children.[31] Although Herbert Herman was

---

[27] For more information, see Zelizer, *Pricing the Priceless Child.*

[28] Interview with Deloris (Boschee) Zimmerman.

[29] Interviews with August Heyne, Reinhold Klein, and Bill Wald.

[30] Interview with Bill Wald.

[31] Interviews with Adam Boschee, Hartmann, Jacob Helbling, Herbert Herman, Herauf, and Geraldine (Dukart) Keen.

only five years old when his grandmother died, he vividly remembers her funeral. Herbert's parents, Jacob and Otillie (Roloff), did not let him attend the funeral services. It was simply too cold outside, Herbert insists. Nonetheless, he was able to watch the service from the warmth of his home, through the kitchen window, which faced the Gackle Cemetery.[32]

For most children, the death of a grandparent was their first experience losing a loved one. Many narrators reminisce at length about their grandparents' illnesses, accidents, deaths, and funerals. For German-Russians, their childhood memories may fade over time, but their memories of losing a grandparent remain clear.[33] Marvin L. Hartmann of Delmont, South Dakota, remembers the funeral traditions were unnerving. He vividly recalls his grandmother Louisa (Hildebrand) Hartmann's funeral. To him, death seemed quite natural, but having his grandmother's body and coffin in his house on the day of the funeral was not. Before the services began, Marvin agreed to help his mother dust the folding chairs. Once he realized he was alone in a room with the casket, Marvin, in his own words, "got the heck out of there." Yet experiences like Marvin's, despite their emotional intensity, underscored the broader cultural values and realities German-Russian children learned from their grandparents.[34]

Indeed, the significance of the grandparent-grandchild relationship extended beyond individual memories or moments of loss. For German-Russian descendants, these intergenerational bonds provided children with a meaningful connection to their cultural heritage, ethnic identity, and family history. Grandparents actively transmitted traditional stories and customs, demonstrated clear social expectations and gender roles, and exposed children to fundamental realities like life, hardship, and death. By bridging past and present, these relationships offered stability, continuity, and a deeper understanding of what it meant to grow up German-Russian on the northern plains. The grandparent-grandchild relationship thus helped maintain ethnic identity and cultural cohesion within this community.[35]

While the grandparent-grandchild relationship played a vital role in the lives of German-Russian descendants, the parent-child relationship—in dis-

---

[32] Interview with Herman.

[33] Interviews with Adam Boschee, Hartmann, Jacob Helbling, Herman, and Keen.

[34] Interview with Hartmann.

[35] Interviews with Bitz, Adam Boschee, Alice Buck, Drefs, Raymond Erbele, Clara Ferner, Graf, Hartmann, Jacob Helbling, Stanley Helbling, Herauf, Herman, August Heyne Keen, Reinhold Klein, Klundt, Long, LeRoy Presler, Sehn, Bill Wald, and Deloris Zimmerman.

tinct and practical ways—shaped ethnic identity. Nearly all German-Russian descendants had direct relationships with their parents, although in rare cases—such as when a parent died young or was otherwise absent—the relationship was experienced primarily through memory or stories told by other family members. These parent-child relationships provided children with a practical education in family expectations, responsibilities, ethnic values, gender roles, and community standards. Unlike grandparents, who served primarily as role models and keepers of tradition, parents typically assumed the role of enforcer. They implemented discipline, taught children that actions have consequences, and guided their children's behavior to fit the expectations of their ethnic community. Whether through direct interaction or family memory, parents influenced their children's identities consciously and deliberately, actively working to preserve their German-Russian heritage while adapting it to their daily realities. Thus, although virtually every child had parents, the intensity or presence of this relationship varied, making it significant but experienced differently among families. .[36]

Through strong parent-child relationships, the second- and third-generation German-Russians learned clear expectations about work and gender roles, a topic discussed further in the upcoming chapter on chores and responsibilities.[37] Esther (Haerter) Drefs, from Venstil, Montana, recalls learning to work hard from her parents. Although her father was not exactly pleased that his daughters had to labor in the fields, Esther regularly help him outdoors and remembers those days fondly. However, reflecting the traditional gender norms held by many German-Russian families at the time, Esther's father considered certain tasks, like sewing, cooking, gardening, or household chores, to be "more gender appropriate" for girls, reserving heavier farm labor and fieldwork as predominantly male duties. Esther thus learned both types of skills—traditional masculine labor in the fields and tasks culturally designated for women at home—and recognized her father's ambivalence toward her involvement in typically male tasks. Esther's memories suggest that her family's discipline and

---

[36] Interviews with Axtmann, Clarence Bauman, Pauline "Polly" (Wock) Berger, Alice (Woehl) Buerkley, Theresa (Meier) Eissinger, Anna (Sturn) Doll, Magdalena (Staudinger) Hauck, Irene (Bender) Schauer, Graf, Clara Ferner, Edwin "Emil" Dewald, Leona (Kuhn) Hoff, Philip Lang, Marjorie (Miller) Flegel, Edna Lu Ella (Weisz) Buck, Drefs, Emil Frank, Wilma (Beck) Haupt, and Alma (Janke) Schott.

[37] Interviews with Polly Berger, Edna Lu Ella (Weisz) Buck, Clara Ferner, Flegel, Magdalena Hauck, and Jacob Helbling.

teachings not only prepared her practically but also reinforced cultural beliefs about gender-appropriate behavior within their ethnic community.[38]

Edwin "Emil" Dewald from Napoleon, North Dakota, shares his story about developing a German-Russian work ethic. Emil recalls having a solid relationship with his father, Jacob, whom he remembers as a hardworking man. Jacob was always busy on the farm—plowing, seeding, and threshing. Emil witnessed this firsthand, as Jacob had his son laboring alongside him most days. When Emil was graduating eighth grade, his father died while working in the fields. Jacob was willing to sacrifice everything, including his life, to care for his family. For Emil, the death of his father embodied the German-Russian ethnic identity. It personified the traditional German-Russian story—a story of working hard and sacrifice.[39]

Discipline, as Dakota Memories narrators insist, was the key to their parents' parental successes. In other words, German-Russian parents from the early to mid-twentieth century were strict, but their children credit their parents' discipline for instilling them with morals and a strong work ethic.[40] Philip Lang, born in Tramping Lake, Saskatchewan, remembers growing up in a home with strict parents. Steven and Katherine (Frohlich), as Philip recalls, were great parents. Yet, he insists his mother ruled with an "iron fist." In the Lang house, children were supposed to be seen but not heard. Philip remembers getting the switch for talking back on a few occasions. Nonetheless, Philip attests their discipline taught him to be a respectful, responsible German-Russian individual.[41]

Through everyday interactions, parents actively passed on practical cultural skills and traditions, as demonstrated by Clarence Bauman, from Java, South Dakota, in his oral history. Clarence recalls learning about traditional German-Russian dress, music, and language from his parents, Fred G. and Rosina (Fiechtner) Bauman. He remembers listening for hours to his father play the organ, while his mother sang songs in German. His father, Clarence points out, was a self-taught musician. Clarence recalls his father deeply valued music. Fred G. made sure to teach each of his children to play at least the basics on the

---

[38] Interview with Drefs.

[39] Interview with Edwin "Emil" Dewald.

[40] Interviews with Henrietta (Gall) Fiechtner, Flegel, Frank, Graf, Magdalena Hauck, Paul A. Hauck, Louis Helfrich, Herauf, and Hoff.

[41] Interview with Philip Lang.

organ and violin. Clarence also credits his parents for teaching him the German language, as that was the only language the Baumans spoke at home.[42]

The parent-child relationship exposed German-Russian children to not only their ethnic heritage but also their socially accepted gender roles.[43] Pauline "Polly" (Wock) Berger of Stark County, North Dakota, recalls clear divisions of labor in her household. Girls primarily handled domestic chores, including baking, gardening, and sewing alongside their mothers. Polly vividly remembers washing clothes by hand, particularly cloth diapers. This task involved soaking soiled diapers, vigorously scrubbing each one on a washboard using harsh, homemade soap, and rinsing repeatedly in cold water. Afterward, each diaper had to be carefully wrung out and hung on the clothesline, regardless of weather conditions. Polly also shared memories of gardening and sewing with her mother. Yet, Polly acknowledges a common reality in many German-Russian families: while daughters often assisted with traditionally male tasks like fieldwork, sons less frequently engaged in domestic chores. This subtle flexibility highlights both the clarity and occasional fluidity of gender expectations within these communities.[44]

Nicholas F. Axtmann, born in Tuscarora Township, North Dakota, remembers learning about his gender responsibilities from his father, Frank J. As a young lad, Nicholas spent a great deal of time working with his father in the fields. He remembers there was no messing around when his parents were near, because they were both strict. Nicholas always knew his father's true desire was to pass the farm on to his son. This plan was the socially accepted method of inheritance in a German-Russian family. At that time, gender determined inheritance practices. German-Russian families reserved the land for male family members.[45]

As Dakota Memories narrators attest, the parent-child relationship also introduced children to social expectations.[46] Alma (Janke) Schott, who grew up near the small town of Gackle, North Dakota, remembers learning about society's expectations for German-Russian girls from both her father and her mother. For example, as Alma approached the completion of the eighth grade

---

[42] Interview with Clarence Bauman.
[43] Interviews with Buerkley, Eissinger, Haupt, Jacob Helbling, Helfrich, Herauf, and Keen.
[44] Interview with Polly Berger.
[45] Interview with Axtmann.
[46] Interviews with Edna Buck and Doll.

Image 3.2. The Nill family photograph (back row, left to right): Ann (Nill) Anderson and Ruth (Nill) Janie; (middle row, left to right) Christian, Ione, and Mary; (front row, left to right): Edgar and Roy F., *circa* 1932–1933, illustrating the pride, stability, and importance of formal portraits in German-Russian family life. *Photograph from the Roy F. Nill Collection, GRHC.*

in a one-room country school, she was optimistic about continuing her education in high school. Her father, however, had other plans. "You don't need any [high school]," Alma remembers her father saying, "[Y]ou're going to be a housewife anyway." Alma was quite disappointed, but she understood her traditional place in society.[47]

The parent-child relationship exposed German-Russian children not only to ethnic heritage but also to religious practices and beliefs. As the narrators report, religion was central to their upbringing. These narrators were Protestant (primarily Lutheran, Methodist, or Baptist) and Catholic.[48] Irene (Bender) Schauer was born in Danzig, North Dakota, and grew up in a Baptist home during the 1930s. She remembers her parents rigidly following church teachings, such as prohibitions against drinking alcohol, dancing, or smoking. Such

[47] Interview with Schott.

[48] Interviews with Magdalena (Staudinger) Hauck and Paul A. Hauck.

behaviors were socially acceptable—and even expected—within their larger German-Russian Baptist community, reinforcing shared cultural norms and fostering a sense of communal identity. For Germans from Russia, religious practices were not just personal or family matters; they structured social expectations, guided community interactions, and defined what was considered morally appropriate behavior. While specific expectations varied somewhat across different denominations, the broader principles of moral discipline, modesty, and moderation were consistently upheld across both Protestant and Catholic German-Russian communities on the northern plains.[49]

This intergenerational relationship played a significant role in the identity formation of nearly all German-Russian descendants. Through the parent-child relationship, German-Russian children learned about their ethnic history and heritage, the importance of a strong work ethic, their gender roles and responsibilities, as well as their society's expectations. Parents not only modeled appropriate behaviors for their children; they also served as disciplinarians. Hands-on training was central to this process; children learned directly from their parents by working alongside them in daily tasks. Examples included boys learning fieldwork skills such as plowing, planting, harvesting, and caring for livestock from their fathers, while girls often worked closely with their mothers to master household skills like baking bread, sewing clothes, tending gardens, and preserving food for winter. Through such practical, hands-on experiences, parents ensured their children developed and maintained a strong German-Russian identity.[50]

The intragenerational relationships of German-Russian descendants also played a significant role in identity formation. This relationship allowed children to forge their own, distinct identities. While the intergenerational relationships kept children grounded in their history and heritage, the generational relationships allowed children to create new experiences. These relationships provided children with an outlet for sharing knowledge, exploring the world, and enjoying life. Unlike the immigrant first generation who struggled to adapt to a new land and maintain their original culture, second- and third-generation German-Russians grew up on the northern plains as their homeland. Consequently, their generational relationships reflected a more secure sense of belong-

[49] Irene (Bender) Schauer, interview.

[50] Interviews with Axtman, Clarence Bauman, Polly Berger, Buerkley, Alice Buck, Dewald, Drefs, Doll, Eissinger, Ferner, Fiechtner, Flegel, Frank, Graf, Magdalena Hauck, Paul Hauck, Haupt, Jacob Helbling, Philip Lang, and Schott.

ing, allowing them greater freedom and confidence to modify or expand their ethnic traditions. Ultimately, it was the generational relationship that allowed German-Russian descendants to create new identities—or at least reconstruct existing ones—suited to their unique, non-immigrant context.[51]

As German-Russian families were rather large, it was common for older siblings to look after younger siblings.[52] For instance, in the Boschee home Deloris (Boschee) Zimmerman was responsible for watching after her three younger siblings, Shirley, Ruben, and Adam. According to Adam, Deloris "didn't spare the rod!" He shares a story about the time the three youngest Boschee children (including himself) went skinny-dipping in the creek with a neighbor girl. When Deloris discovered the children swimming without clothes, she gathered up all the clothes and hustled the swimmers out of the lake. Adam recalls having to walk two hundred to three hundred yards back to the house without clothes or shoes, exposing his feet to sharp prairie hazards like stickers (small, prickly burrs common to the northern plains) and rocks.[53] When asked about this incident during her 2005 oral history, Deloris cunningly smiled and quietly chuckled. While Adam describes this as discipline, Deloris insinuates it might have been a small prank amongst siblings. Either way, Deloris was sharing her wisdom with her younger brother, also known as teaching him a lesson.[54]

Dakota Memories narrators also remember their parents leaving them unsupervised for a few hours or a few days, giving them the freedom to create their own experiences. Typically, parents left older children in charge of the younger ones. However, if the children were close in age, parents expected them to look out for each other. Born and reared near Strasburg, North Dakota, in the 1930s, Armelia "Millie" (Ibach) Nagel remembers her parents—Joseph and Magdalin (Senger) Ibach—leaving her home alone with her younger siblings on multiple occasions. Given that there were eleven children in the Ibach home, it was nearly impossible for them all to attend church at the same time. During her July 2007 oral history, Millie reminisces about an incident that occurred one Sunday when her parents left her home with two of the younger children.

---

[51] For more information on the significance of sibling relationships in regard to identity formation, see Atkins, *We Grew Up Together*; Interviews with Backfish, Marilyn Jeanette (Sauer) Bauman, Adam Boschee, Buerkley, Edna Buck, Dewald, Doll, interview; Marvin G. Fiechtner, Robert R. Gab, Helfrich, Hoff, Anne (Frehlich) Hoffart, Michael Jahner, Lapp, and Elda (Schultz) Rasch.

[52] Interviews with Buerkley and Doll.

[53] Interview with Adam Boschee.

[54] Interview with Deloris Zimmerman.

Millie is unclear as to what exactly, but something had spooked the Ibach children. She does not recall precisely how many siblings joined her or exactly how long they stayed hidden, but she clearly remembers gathering the frightened children, along with the family's Shetland pony and hiding in the outhouse. In an attempt to protect her younger siblings, they remained concealed until their parents returned from church. Millie adds that the Shetland pony, of course, was tied just outside the door of the outhouse.[55]

This sibling relationship also provided children with an outlet to share in the experiences of exploration.[56] For instance, Dakota Memories narrator August Heyne, born in Grant County, North Dakota, remembers exploring the countryside with his siblings: Arthur, Edwin, Emil, and occasionally Loretta. He remembers swimming and hunting with them quite often. They learned to swim in a nearby river. According to August, they learned by spending a good deal of time playing in the river. August reports that the Heyne boys also did a good deal of hunting for rabbits, pheasants, and deer, and trapping badgers, weasels, and skunks. Through these activities, the Heyne boys became regional experts. At the same time, they were learning about physical surroundings—the environment, land, and animals—that were foreign to their immigrant ancestors. Together, the Heyne boys became German-Russian on the northern plains (as did other German-Russian descendants through their adventures and explorations).[57]

Intragenerational relationships provided countless German-Russians with opportunities for socializing. Narrators of the Dakota Memories reminisce about playing make-believe, ball, marbles, race games, listening to music, playing instruments, and roaming fields with their siblings.[58] While these shared activities are explored in greater detail in the chapter devoted to play and pastimes, it is important to note here that sibling interactions served a key role in reinforcing social connections and collective identities. Esther (Haerter) Drefs remembers being close to her siblings who survived infancy. She remembers they had fun together, playing all sorts of games. She insists, however, there was never any real competition or sides. It was just for fun.[59]

---

[55] Interview with Armelia "Millie" (Ibach) Nagel.

[56] Interviews with August Heyne, Drefs, Marvin Fiechtner, and Gilbert Ost.

[57] Interview with August Heyne.

[58] Interviews with Backfish, Clarence Bauman, Edna Buck, Flegel, Paul Hauck, Victoria (Schaff) Hecker, and Helfrich.

[59] Interview with Drefs.

Image 3.3. Pauline "Polly" (Wock) Berger's sister Wilfreda "Willie" (Wock) Herauf enjoyed helping their brother Willard with the "de-skunking" of skunks, an example of the resilience, humor, and cooperative spirit that shaped German-Russian childhood. *Photograph from the Pauline "Polly" (Wock) Barger Collection, GRHC.*

However, teasing and hazing were also common among German-Russian siblings. On the plains, children experienced plenty of freedom from adult supervision, allowing them to engage in several childlike activities, such as playing pranks or tricks on siblings.[60] Edwin "Emil" Dewald vividly remembers his brother and sister playing tricks on him all the time. Arthur, Alvin, and Elsie, Emil's siblings, loved having fun at Emil's expense. Emil shares two stories to corroborate his assertions—the sticky doorknob incident and the smelly delivery story. One time, as Emil tells it, Alvin and Elsie decided to make candy, something their parents forbad unless under their supervision. When their parents returned home and found a sticky doorknob, Alvin and Elsie did not hesitate to place the blame squarely on Emil's shoulders. On another occasion, Alvin and Elsie decided it would be funny to place a fresh pile of smelly cow

[60] Interviews with Dewald, Gab, and Rasch.

manure in the mailbox. Since Emil was the youngest, it was his responsibility to retrieve the mail. Emil, who was also the shortest in the family, could not see into the mailbox, and when he reached his hand into the mailbox, he found only a gooey pile of poop.[61]

Elda (Schultz) Rasch, from Fredonia, North Dakota, also remembers her siblings' incessant teasing. Her older brother, Lehnard, was the primary instigator. Thirteen years her senior, Lehnard pulled off an assortment of pranks. Elda remembers him throwing her into a bed, covering her up with the blankets, and telling her to stay. He would then exit the room, leaving young Elda hidden in a pile of blankets in a silent room. She remembers as a little girl this terrified her. Elda never explicitly states why this terrified her, perhaps assuming listeners would instinctively understand the fear a small child might feel when suddenly isolated, unable to see clearly, or uncertain about when—or if—her brother would return. Lehnard also found it amusing to put Elda on the back of a young calf and to run it around in circles. Eventually, as Elda recalls, she would get dizzy and fall off the calf. It just so happened that she usually fell into a pile of fresh manure. Thinking back on her childhood, she cannot help but smile at all the fun Lehnard had at her expense.[62]

As the Dakota Memories narrations demonstrate, children not only had fun by pulling small pranks and tricks on their siblings, but they also had fun by creating mischief. Narrators of this project, however, typically gloss over issues of mischief, likely because openly acknowledging such behavior contradicts deeply held German-Russian values emphasizing discipline, obedience, and community approval. Admitting to disobedience or disruptive behavior might conflict with how narrators want to be perceived—both individually and as representatives of their ethnic community. Fortunately, the oral history teams have, on occasion, broken past this collective memory barrier and tapped into deep, individual memories of childhood misbehavior. These deeper memories reveal that these German-Russian descendants did in fact misbehave.[63]

Selma (Job) Lapp, from McPherson County in North Dakota, remembers finding herself in some mischief as a young girl. When she was about six or seven, she decided that it would be fun to play with her cousins in the fami-

[61] Interview with Dewald.

[62] Interview with Rasch.

[63] Interviews with Clarence Bauman, Doll, Henrietta Fiechtner, Flegel, Laura (Schulz) Hellmuth, Herauf, Anne Hoffart, Hoff, and Michael Jahner.

ly's parked car, which she knew was off limits. Someone accidently pulled the wrong lever, sending the car straight toward the outhouse. Selma recalls frantically screaming for help. Fortunately, someone heard Selma's cries and dashed outside to stop the car before it hit the outhouse. She remembers getting a good scolding for this incident.[64]

Marvin G. Fiechtner, who was born in Wishek, North Dakota, remembers stirring up his own mischief. As an only child, he had to create his shared experiences with his cousins or other town children. He really enjoyed hanging around one fellow, the son of the town's grocer, who was a few years his senior. Marvin recalls this boy always had easy access to tobacco. Actually, Marvin insinuates that this boy was stealing it from his father's grocery store. With the tobacco in hand, Marvin and his friends headed to the outskirts of town to smoke. In Marvin's own words, they smoked "like a bunch of Indians sitting around a campfire," reflecting common perceptions German-Russian children had about Native Americans—perceptions shaped by popular childhood games and the stereotypes of the era. While playful in intention, such descriptions inadvertently highlight how ethnic communities define themselves, in part, by their ideas about other groups.

Marvin emphasizes that at first, he was not a willing participant, but eventually he succumbed to the peer pressure. He remembers feeling horrible after his first couple of times smoking. Yet, he eventually developed a tolerance for it. During this story, Marvin even demonstrates the process of rolling cigarettes, including the best method for licking the paper. Through such memories, narrators like Marvin provide subtle insights into the broader cultural environment in which their German-Russian identities were formed.[65]

Growing up on the northern plains, German-Russian descendants had new experiences they shared with others of their generation, especially siblings and cousins. Unlike their immigrant grandparents, who struggled with displacement, harsh frontier conditions, isolation, language barriers, and limited resources, the second- and third-generation German-Russians grew up more comfortably rooted in their American and Canadian communities. Through in-

---

[64] Interview with Lapp.

[65] Interview with Marvin Fiechtner.

tragenerational relationships, these children shared enjoyable aspects of life on the plains, such as games, adventures, and mischief. While the first generation primarily emphasized discipline, survival, and maintaining cultural continuity amid difficult circumstances, later generations experienced greater freedom to explore, adapt, and even challenge traditional expectations. These new, shared experiences on the northern plains allowed the second- and third-generation German-Russians to reshape their ethnic narrative and identity. Thereby, the German-Russian story evolved beyond a narrative solely defined by persecution, migration, and hardship; it became one of adaptation, playful exploration, and accomplishment. Incorporating these childhood memories brought to light by deep interviewing offers a fuller understanding of how ethnic identities change over generations, balancing historical struggles with newer narratives of growth, joy, and acheivement.[66]

The social construction of family is a central theme for the German-Russian narrations collected by the Dakota Memories project. While kinship refers specifically to biological or genetic connections, the German-Russian descendants emphasize family as something both socially and culturally defined – built through shared experiences, emotional bonds, and collective identity. They proudly highlight their ability to sustain familial ties across multiple generations and international boundaries. Narrators regularly discuss their relationships with grandparents, parents, siblings, and extended family, all of which helped shaped their identities and provided foundation for their stories.

More specifically, intergenerational relationships allowed German-Russian descendants to explore their ethnic history, heritage, and cultural practices. Grandparents served as reflective storytellers who passed down memories, traditions, and societal expectations. Parents, however, provided practical instruction and hands-on training in tasks related to daily life, teaching children about work, gender roles, social responsibilities, and religious obligations. In this way, each type of relationship uniquely contributed to the formation of the second- and third-generation German-Russian identity.

The intragenerational relationship of German-Russian descendants was also quite significant for the second and third generations. Through this re-

[66] Interviews with Backfish, Marilyn Bauman, Adam Boschee, Buerkley, Edna Buck, Dewald, Drefs, Doll, Fiechtner, Flegel, Gab, Paul Hauck, Hecker, Helfrich, Michael Jahner, Selma (Job) Lapp, Armelia "Millie" Nagel, Elda (Schulz) Rasch, and Deloris (Boschee) Zimmerman.

lationship, German-Russian descendants forged a new German-Russian story, a more complex German-Russian story. Memories of freedom, pastimes, and mischief directly contradict this ethnic group's established collective narrative: life was hard. Memories of their familial relationships, especially their intragenerational relationships, reveal that narrators of the Dakota Memories believe they had happy childhoods despite some hardships.

This new research provides the first in-depth analysis of family relationships among the second- and third-generation Germans from Russia. The analysis reveals that the experiences of German-Russian descendants on the northern plains differed greatly from their ancestors. While the first generation primarily defined their identities through shared struggles on an unfamiliar frontier, later generations constructed their identities from a position of greater stability and familiarity with their homeland. Unlike their grandparents, the second and third generations grew up in established communities, allowing them more freedom to adapt, expand, and even challenge traditional expectations. Thus, while the identity of the German-Russian descendants remained rooted in the traditions and memories of their ancestors, it was continuously reshaped by their own new and distinctly American or Canadian experiences. These German-Russian descendants, for the most part, express no bitterness during their interviews, although they grew up on the plains during the treacherous times of the Great Depression. They embrace their hard times and sufferings by making it only part of their narratives. They found value in their heritage, yet, at the same time, joy in their memories. In addition, this research also supports the ideas presented in the existing historiography of childhood in the West (works by West, Riney-Kehrberg, and Holt) and the current literature on family relationships (works by Barranti, Pollack, and Atkins). It reaffirms the importance of familial relationships in the formation of identity. Furthermore, it also provides a new category of analysis for family and childhood historians of the American West: ethnicity.[67]

Selma (Job) Lapp's oral illustrates the significance of familial relationships—both intergenerational and intragenerational—for Dakota Memories

---

[67] For more information, see Shopes, "Using Oral History for a Family History Project"; West, *Growing up with the Country*; Riney-Kehrberg, *Childhood on the Farm*; Holt, *Children of the Western Plains*; Thompson, *Voice of the Past*; Barranti, "The Grandparent-Grandchild Relationship"; Linda A. Pollock, *Forgotten Children: Parent-Child Relations from 1500 – 1900* (New York: Cambridge University Press, 1983); Atkins, *We Grew Up Together.*

narrators. Although Selma never had a personal relationship with her great-grandparents, her family folklore has allowed her to develop a meaningful connection to them, deepening her sense of ethnic identity. . She did, however, have a personal relationship with her grandparents, whom she remembers warmly, recalling her grandfather reading aloud from German newspapers and her grandmother preparing delicious dough dishes.. Selma also spent a good deal of time reminisces fondly about her parents, emphasizing how their stories, actions, and disciplined approach taught her to value hard work. Additionally, she recalls enjoyable generational interactions with siblings and cousins such as the mischievous runaway car incident recounted earlier in this chapter. Her experiences differ substantially from those of her ancestors: while the immigrant generation primarily recalled hardships related to migration, displacement, and adaptation, Selma's narrative emphasizes joy, playful exploration, and a comfortable sense of belonging on the northern plains. Thus, through thesefamilial relationships, Selma constructed a new, uniquely American story rooted in her ethnic heritage, yet shaped by her distinct experiences. In this way, her family story becomes "more than a photo," as Dakota Memories narrator Calvin K. Fercho describes—it is a living narrative blending past and present, tradition and innovation..[68]

[68] Selma (Job) Lapp, interview; Calvin K. Fercho, interview.

CHAPTER 4

Work Makes Life Sweet: Chores and Responsibilities

"We all had our chores," says Marilyn Jeanette (Sauer) Bauman of Java, South Dakota, "things we had to do." By the time she was six, Marilyn, had started milking cows, and by eight, she was overseeing her family's entire milking operation. That meant ensuring the younger Sauer children completed their share of the work before heading to school in the morning or to bed in the evening. In addition to managing the milking and separating processes, Marilyn had another daily task: gathering fuel for the kitchen stove. This fuel came in the form of dried, and as she states "smelly," cow chips, which she collected by hand. Looking back, she admits the memory of handling them without gloves makes her cringe, but at the time, it was simply another necessary chore.[1]

In 2007, Marilyn and her husband Clarence Bauman—also from Java—participated in the Dakota Memories project together, as well as individually. They sought to preserve and shape their ethnic narrative.[2] As a key element in this exercise of memory and identity, countless narrators, like the Baumans, reminisce about milking cows, separating milk, hauling hay, picking rocks, ironing laundry, and baking bread. Narrators of the Dakota Memories initially emphasize their dreary chores and stern responsibilities in their oral histories, but after deeper inquiry they reveal the complexities of their childhoods.[3] These

---

[1] Interview with Marilyn Jeanette (Sauer) Bauman.

[2] Interview with Marilyn Bauman and Clarence Bauman.

[3] Interviews with Elizabeth Barbara (Rissling) Lang, Pauline "Polly" (Wock) Berger, Edna Lu Ella (Weisz) Buck, Robert R. Gab, Christina Rose Marie (Bast) Krismer, Irene (Bender) Schauer, and Wendelyn Vetter.

deep memories shed light on the significance of childhood labor—not just as toil, but as a source of practical experience, socialization, gender roles, economic necessity, personal satisfactions, and risks.

The collective memory of Germans from Russia is one of hard work; children labored from a young age, often in demanding conditions. Yet, their individual and deep memories provide nuance for this narrative, revealing how work also initiated opportunities and lasting experiences. Resilience is central to their identity, but it is not the sole defining trait. For many first-generation immigrants, perseverance through hardship became an essential marker of what it meant to be German-Russian. However, the second and third generations, while still deeply valuing resilience, also define their identity through cultural traditions, familial bonds, faith, and community. As the Dakota Memories narrations reveal, chores and responsibilities played a key role in intertwining the bitter and the sweet aspects of childhood, shaping both their ethnic memory and the ongoing reconstruction of the German-Russian story.

Only in the last thirty years have historians examined rural children as historical actors, recognizing them as active participants in shaping their communities rather than passive bystanders in history. This new, rich, historiography reveals that chores and responsibilities played a significant and complex role in the lives of children in the West. Some of the early works addressing rural childhood gloss over the hard physical labor and taxing responsibilities children faced at young ages. Other works argue that childhood labor on the frontier often crossed the line from necessary work into exploitation, with children enduring relentless toil, limited schooling, and harsh discipline. Historians of late have provided a more even-handed analysis on childhood chores and responsibilities, such as can be found in Marilyn Irvin Holt's *Children of the Western Plains* and Pamela Riney-Kehrberg's *Childhood on the Farm*, which acknowledge the hardships and the opportunities childhood labor provided.[4]

Chores and responsibilities were not only demanding for children, but highly educational, serving as hands-on training for adulthood. Daily tasks provided practical experience in essential skills such as animal husbandry, crop management, food preservation, and financial literacy. These responsibilities reinforced work ethics and family duty, ensuring that children were prepared for the roles they were expected to assume as adults. While many tasks were

---

[4] For the former historical works described here and in previous chapters, refer to authors Elliott West (*Growing Up with the Country*) and Elizabeth Hampsten *Settlers' Children*).

divided along gender lines, children often stepped into whatever role was needed based on immediate farm demands. Beyond survival, these chores also reinforced cultural identity, as children learned traditional farming techniques, household management, and food preparation methods passed down through generations.[5] Though exhausting, this labor contributed to a strong work ethic and a deep sense of responsibility—qualities that German-Russian families on the northern plains valued highly.

Traditionally, chores reflected socially accepted gender roles, with boys expected to take on physically demanding outdoor work while girls managed household duties such as cooking, cleaning, and childcare. These early responsibilities not only prepared children for their expected adult roles but also reinforced broader societal expectations about masculinity and femininity. However, the realities of farm life often demanded flexibility. When labor shortages arose or immediate needs dictated, children adapted, regardless of gender norms. Though parents may have preferred a structured division of labor, economic survival frequently took precedence over tradition.[6]

Historians acknowledge necessity often dictated work roles more than social expectations. Holt in particular argues that who performed which tasks was largely a matter of practicality rather than rigid gender norms. In German-Russian families on the northern plains, this meant daughters might plow fields or tend livestock, while sons could be found baking bread or caring for younger siblings. These blurred lines of labor not only ensured the family's survival but also fostered a strong sense of adaptability and resilience among farm children. Work was more than an obligation. It was an integral part of shaping individual character and collective identity.

As Viviana A. Zelizer discusses in her foundational work, *Pricing the Priceless Child*, late nineteenth and early twentieth centuries saw a major shift in how children were valued. No longer seen primarily as economic contributors, they were increasingly viewed as emotionally precious dependents whose primary role was to learn and grow rather than work.[7] This change, however, was largely a middle-class, urban ideal. For many rural families, especially on the northern plains, children remained essential contributors to household and agricultural labor. As historian Elliott West observes, most families "could not

---

[5] Riney-Kehrberg, *Childhood on the Farm*, 51.

[6] Riney-Kehrberg, *Childhood on the Farm*, 51; West, *Growing Up with the Country*, 75.

[7] Zelizer, *Pricing the Priceless Child*, 3.

afford an ideal child"—meaning that the expectation of a childhood free from labor was simply not realistic for those who relied on every family member to ensure survival.[8] While public perception of childhood evolved, economic necessity often outweighed societal expectations in farm communities."

For rural families, especially Germans from Russia, chores were more than just labor; they were a value-added investment in a child's development. Hard work was not only expected but seen as a defining characteristic of identity, instilling discipline, resilience, and a sense of purpose from an early age. Yet, beyond necessity, work was not always a burden. Many children found ways to take pride in their tasks, enjoy the satisfaction of accomplishment, and even have fun in the process. Whether it was caring for animals, exploring the land while gathering materials, or turning tasks into games, work often blended with play. Chores also provided opportunities for socialization, especially during communal activities like threshing, when extended families and neighbors came together to share both the labor and the experience. "Helping out," as child labor was commonly described, was not just about survival—it was a way for children to engage with their families, their communities, and the land itself.[9]

While chores provided children with valuable skills and social connections, they also came with significant risks. Farm work was physically demanding, often requiring children to handle livestock, heavy machinery, and sharp tools from a young age. Without strict supervision, the line between routine tasks and hazardous situations blurred quickly. Children's work had the "sheer danger and capacity to hurt, cripple, or even kill young workers," states historian Riney-Kehrberg.[10] Children were not always careful, and accidents were quite common. Farm children, starting at a young age, were typically in close proximity to a number of domestic animals and implements. Once children became familiar with the animals and equipment, it was quite common for them to work unsupervised. Farm work was hard and dangerous.[11] In fact, one childhood historian, Hampsten, even insinuates that child labor on farms was child neglect, if not abuse.[12] Most narrators, however, neither claim nor insinuate that their experiences with child labor were abusive, viewing their

---

[8] West, *Growing Up with the Country*, 74.

[9] Riney-Kehrberg, *Childhood on the Farm*, 36–60; West, *Growing up with the Country*, 74.

[10] Riney-Kehrberg, *Childhood on the Farm*, 56.

[11] Riney-Kehrberg, *Childhood on the Farm*, 36-60.

[12] Hampsten, *Settlers' Children*, 13–34.

work instead as a necessary and expected part of their familial role. Dakota Memories narrators would agree with Riney-Kehrberg, that childhood chores and responsibilities were strenuous, laborious, and at times dangerous. Yet, for German-Russian families, these experiences were more than just work; they were deeply tied to ethnic identity. Hardship was not simply endured—it was embraced as part of what it meant to be German-Russian. Narrators, however, neither claim nor insinuate that child labor was abusive. Instead, they describe a life where work was relentless but also fulfilling, shaping their resilience and self-sufficiency. Several reminisce about beginning chores at three to five years old, recalling how their parents introduced them to labor at an early age by having them observe and assist with simple tasks. Before they were expected to contribute meaningfully, they were immersed in a culture where work was a defining aspect of daily life, blending duty with shared experiences. The Dakota Memories oral histories reveal chores and responsibilities were present in all stages of childhood, infancy (birth to two years of age), where children were carried along to the fields or kitchens, to early childhood (two to five years of age), when they began helping with small tasks, to childhood (five to twelve years of age), when they took on more physically demanding labor. Despite the difficulties, narrators frequently intertwine memories of hard work with stories of camaraderie, play, and family bonding, reinforcing the notion that a tough life did not preclude joy.

With few memories of their infancy, some narrators rely on folklore, oral tradition, or assumptions to fill in the blanks. Some restate stories handed down to them from their parents or grandparents, others share childhood observations of their younger siblings. According to the narratives, the immigrant generation exposed its children to hard work before they were old enough to walk. Narrators point out that as infants they were close by as their families worked long hours, day in and day out. They observed as their family members attended to the chores. Narrators of this project believe that this early exposure to chores was not only practical but also educational. If infants were close enough to observe the daily routines, they were close enough for family members to watch over them. Furthermore, by observing the daily life of their family, infants were learning about the chores they would one day assume.

Eileen M. (Burlack) Ketterling, from Burnstad, North Dakota, remembers her younger siblings being exposed to chores and farm work at an early age. When she was nine, she was given the responsibility of caring for her younger

Image 4.1. Esther (Kauk) Schlecht's family haying, *circa* 1933, showing the collective effort and pride in shared farm work that defined German-Russian family life. Pictured (left to right): Lawrence Kauk, Barney Kauk, Esther (Kauk) Schlecht, and Magdelina Kauk (Esther's mother). *Photograph from the Esther (Kauk) Schlecht Collection, GRHC.*

brother, John—entertaining, changing, and feeding him. As she recounts her memories in her oral history, she vividly describes taking John out to the fields where their mother was laboring so he could be breastfed. Mid-story, however, Eileen becomes visibly embarrassed, likely due to the modesty she was raised with and the sensitive nature of the topic. Quickly shifting focus, she begins reminiscing about the bumpy stroller ride that John had to endure on the way out to the field. Although John was just an infant, his sister was exposing him to a variety of chores, like fieldwork. She also toted him around while feeding the chickens and calves, preparing the noon meals, and gathering the water. For

Image 4.2. Michael Anton Wock with his children, Willard Lawrence and Eugenia Marie (Schmidt), *circa* 1930s, illustrating how fathers passed down skills and values through shared farm life, strengthening family bondss. *Photograph from the Pauline "Polly" (Wock) Barger Collection, GRHC.*

Eileen and many others, chores were not something children grew into; they were a presence in life from the very beginning.[13]

LeRoy Presler, born in Gackle, North Dakota, remembers the harsh realities of early exposure to farm chores. Holding back his emotions, he shares the painful story of his younger brother, Wallace, the youngest of eighteen children, who was just shy of two when tragedy struck. While the older Presler children labored in the fields, the younger ones tended the homestead and cared for the babies. The day Wallace died, he was following his sisters as they carried water to the field when he wandered too close to a protective mare. The horse lashed out, striking him with a fatal kick. He died instantly. Decades later, LeRoy's voice still tightens at the memory—a heartbreaking reminder that for farm children, even the simplest tasks could carry devastating consequences.[14]

Although early childhood memories are unreliable, they are more detailed than memories from infancy, offering glimpses into personal experiences and

[13] Interview with Eileen M. (Burlack) Ketterling.

[14] Interview with LeRoy H. Presler.

identity formation. These recollections frequently take the form of episodic memories—distinct, experience-based recollections tied to emotions and specific events. Rather than vague impressions of the past, these memories allow narrators to reconstruct their early sense of self within the framework of their cultural heritage. Germans from Russia often use these childhood memories to reinforce their collective identity as a resilient, hardworking people who overcame hardship and persecution. However, the Dakota Memories narratives reveal a more nuanced perspective. While early childhood recollections still reflect a strong work ethic, they also highlight the multifaceted nature of chores—beyond labor, they served as lessons in job skills, gender roles, and familial responsibility. These memories capture a full spectrum of experiences, both challenging and rewarding, illustrating how work was not only a necessity but also a formative part of childhood. While only a few narrators of the Dakota Memories have personal recollections of their infant years, many have memories, albeit sketchy ones, of their early childhood years. According to them, this stage marks a shift from simply witnessing family life to actively contributing to it. In childhood history, historians have reframed the experiences of children from passive observers to active participants, engaged with and shaping the narrative. As active participants, they take on small but meaningful roles—helping with chores, learning family traditions, and assisting with daily responsibilities. Rather than existing on the periphery, they are seen in the rhythms of family labor and social interactions. Their contributions, though sometimes dismissed as insignificant, were essential in rural households, where every family member played a role in sustaining the home and farm. This transition from passive presence to active engagement reflects a broader concept in childhood history: children are not just shaped by their circumstances; they also actively shape their own experiences and, in doing so, contribute to the historical narrative of their communities.

For many, this transition into active participation began remarkably early. Narrators assert that children as young as three or four years of age helped with the domestic chores. For instance, Viola (Eslinger) Konrad, from Fredonia, remembers helping with the chores before she was five years old. Viola recalls that one of her first chores was to assist with the milking. Her job was to stand behind the cows and hold their tails to keep them from swatting at the milkers. With flies constantly buzzing, the cows' tails were in near-constant motion, making her role both necessary and challenging.[15]

[15] Interview with Viola (Eslinger) Konrad.

Image 4.3. John, Eileen M. (Burlack) Ketterling's younger brother, helping with the bundles during harvest, reflecting the vital role of children in ensuring the family's success and building a strong work ethic from an early age. *Photograph from the Eileen M. (Burlack) Ketterling Collection, GRHC.*

The chores of children five and under were quite simple. Nevertheless, these simple chores provided children with a practical education and onsite job training. Simple domestic chores taught children the basic skills needed for survival, as well as responsibility. Peter A. Kraft, from Strasburg, North Dakota, remembers being a "modest helper" as a young lad. Peter claims he was his mother's maid. During his early childhood, Peter gathered eggs, cow chips, and kindling. These chores taught him to be a responsible member of his family.[16]

"It was a little harder and harsher when I was growing up," claims Pauline "Polly" (Wock) Berger of Stark County, North Dakota. She insists that children had to start working early because their labor was essential to the household. As a young girl, she was responsible for feeding the chickens, hogs, and cattle—chores that required both attentiveness and strength. During her oral history session, Polly vividly describes the process of bucket-feeding calves, recalling how they eagerly pushed against her, their rough tongue lapping at the milk as they fought to drink. If a child was not careful, their toes could easily be tram-

[16] Interview with Peter A. Kraft.

pled by impatient calves lunging for their share, she notes. Polly remembers her parents warning her to stay alert, but the risk was simply part of the job. The family relied on their livestock for milk, cream, and butter—key sources of food and income. Milking, feeding, and caring for the animals were not menial chores; they were essential tasks that ensured the household's survival and economic stability.[17]

Since these memories are episodic and blurry, many narrators focus solely on recounting the basics, often unable to elaborate due to the natural constraints of early childhood recollection. Rather than detailed stories, they recall simple, repeated actions—tasks that became routine rather than singular events. Louis Helfrich, from Glen Ullin, North Dakota, and Irene (Bender) Schauer from Danzig, for example, remember hauling water at just four years of age. Louis and Irene have very similar memories of their lived experiences, despite living in distinctly different regions of North Dakota (Louis in western North Dakota and Irene in south-central North Dakota). Nonetheless, their memories of hauling water resemble each other. As they recall, water was one of two commodities children were responsible for replenishing (the other, of course, being fuel). In the Helfrich home, Louis remembers hauling water from a nearby creek for his parents. The water, as he recalls, was used for laundry, drinking, and bathing. In the Bender home, Irene also remembers hauling water for domestic use. She concludes her narrative by acknowledging that younger children of course carried less water than older children. Her memory, like Louis's, is not rich in detail but rather reflects the nature of early childhood recollection—structured around routine, repetition, and the small yet significant ways children participated in household survival.[18]

Memories from childhood are more reliable than memories from infancy and early childhood. The memories gleaned from this stage of development provide rich, new insights into previously examined topics. For instance, the Dakota Memories narrations reveal that the second- and third-generation German-Russians have a much more complex childhood than they originally indicated. A number of German-Russians decline to participate in the Dakota Memories because they insist that they do not have *anything interesting to say*, a common sentiment among potential narrators. Yet, Dakota Memories oral historians have successfully documented that German-Russians do have

---

[17] Interview with Polly Berger.

[18] Interviews with Louis Helfrich and Schauer.

a fascinating story, with crucial nuances that only they can tell. For instance, their chores not only made their childhood hard, but also provided them with a practical education, helped their families survive, and made their childhoods memorable.

By the time they were five or six years old, chores were a part of everyday life for German-Russian children. Chores were never-ending, Elizabeth Barbara (Rissling) Lang, from Scott, Saskatchewan, maintains while reminiscing about growing up in a German-Russian farm family. It was impossible for the entire family to enjoy any community celebration, because the cows needed milking, the milk needed separating, and the equipment needed cleaning every day, twice per day. She remembers the monotony, yet importance, of this work. As a young girl, Elizabeth only attended one Canada Day celebration as she was needed at home on the farm. She summed up her sacrifice by saying, "[S]omebody had to stay home," accepting it as an undeniable part of farm life.[19]

The educational benefits of chores included exposure to traditional social roles. Children ages six to twelve years old experienced an increase in responsibilities. Many parents believed working on the farm provided children with the only education they needed. Growing up in Tuscarora Township in Pierce County, North Dakota, Nicholas F. Axtmann received portions of his education on the farm. When he was a young boy, his parents held him out of school quite often during the fall season. He remembers missing anywhere from two to four weeks of school at a time. It was hard to catch up, he recalls. Yet, he concludes, "I did pass!" His parents kept him home from school in part because they needed his help, and in part because the farm chores were providing a practical education.[20]

The skills German-Russian children learned by doing chores prepared them for a life in a rural and agricultural region. Ida (Zimmerman) Koepplin learned about farm life and responsibilities while growing up in Lehr, North Dakota. When she recounts her childhood memories of chores, Ida discusses milking cattle, feeding chickens, "setting" hens—the process of placing broody hens on a clutch of eggs to encourage hatching—shearing sheep, and butchering hogs.[21] Philip Lang remembers learning important skills and traits on the family farm at Tramping Lake, Saskatchewan, as well. He recounts his memories of

---

[19] Interview with Lang.

[20] Interview with Nicolas F. Axtmann.

[21] Interview with Ida (Zimmerman) Koepplin.

milking cows, cleaning barns, harvesting fields, and threshing crops. Cleaning the barn was a long and unpleasant chore, but it was a daily expectation set by his father. After school, it was his job to ensure the barn—home to ten to fifteen horses—was spotless before his father came to inspect it. He remembers using a one-wheeled wheelbarrow to haul manure, carefully piling it up for removal. But sometimes, just before reaching the barn door, the wheelbarrow would tip, spilling its contents back onto the ground. The frustration he felt as a boy is evident as he describes having to reload the manure, knowing that he could not stop until the task was done properly.[22] Chores like these didn't just teach Philip the basics of farm work; they instilled discipline, perseverance, and an understanding that expectations had to be met no matter how frustrating the job.[23]

Chores also helped children adapt to their gender identity at this stage in life. Typically, girls carried out or assisted with the domestic chores, while the boys carried out or assisted with the field labor. In the Ell household, girls did the domestic chores. Clara (Ell) Ferner. From Kronau, Saskatchewan, is one German-Russian who vividly remembers her domestic responsibilities. She insists children did everything. She has memories of cleaning the kitchen, washing the dishes without detergent, and scrubbing the floors with lukewarm water. She also remembers that children in the Ferner house were quite concerned about equity. That is, they made sure no one had more chores than the others. Clara and her sister had an imaginary line in the kitchen that neither she nor her sister could ever cross. They were not going to "help" each other. Stumbling onto this private memory brought joy to Clara, so much so that she could not help but smile.[24]

One chore narrators identify as traditionally in the domestic realm is laundry. Some narrators claim they never helped with the laundry because it was too much work; others, however, remember spending countless hours working alongside their mothers on washday. Those who remember Monday as washday remember it was an "all day affair."[25] The first step, as narrators recall, was hauling water. During the first half of the twentieth century, rural homes on the northern plains lacked running water, and thus children had to fetch water from a nearby creek or well (like Louis Helfrich and Irene [Bender] Schauer did, as

---

[22] Interview with Philip Lang.

[23] Interviews with Lang and Ida (Zimmerman) Koepplin.

[24] Interview with Clara (Ell) Ferner.

[25] Interview with Florence (Mitzel) Engelhart.

previously mentioned).[26] The second step involved heating the water, typically on a coal-burning stove, although it was also common for German-Russians to use dried cow chips or corncobs as fuel to heat stoves. Once the water was warmed, it was time to start washing the clothes. A few narrators remember using old scrub boards, but most remember using motorized washing machines. These washing machines were difficult to start. After completing the wash, narrators recall hanging their clothes out to dry, regardless of the temperature. Thinking back on their childhood, many narrators are still amazed by the sight of their clothes frozen stiff in the bitter cold.[27]

Cooking and baking were other chores typically reserved for girls. Magdalena "Maggie" (Kuntz) Degenstein remembers helping with the baking quite often in her family's Orrin, North Dakota, home. Maggie had to take on adult responsibilities, as her mother was often sick. In addition to looking after her three younger siblings—a role requiring patience, supervision, and caregiving—she was responsible for preparing the meals for the family, a task normally expected of adult women. She recalls baking six to seven loaves of bread, two to three times per week, a crucial responsibility since bread was a staple at every meal.[28] Florence (Mitzel) Engelhart, from Strasburg, North Dakota, also remembers doing a great deal of baking and cooking, describing it as an essential part of her upbringing. She learned from her mother, not just out of necessity, but as part of her cultural education. Many recipes were passed down through generations, including knepfla (small dumplings often served in soups), strudla (rolled dumplings steamed over meat and potatoes), and kuchen (a sweet, custard filled pastry). Florence still cherishes these skills today, continuing to make these dishes as a way of preserving her German-Russian heritage.[29]

Just as girls took on strenuous household responsibilities, boys were expected to handle physically demanding field labor. Balzer Kurtz remembers he was responsible for hauling coal and railroad ties to his Devils Lake, North Dakota, home to fuel the family's fires. Living near the railroad line, his family relied on wood for about 80 percent of their heating fuel, with his father prefer-

---

[26] Interviews with Helfrich and Schauer.

[27] Interviews with Polly Berger, Florence (Mitzel) Engelhart, Magdalena Hauck, Helfrich, Anne Hoffart, Lyla (Pressler) Johnson, Lorraine (Meidinger) Kaseman, Leona (Miller) Klundt, Larry Kraft, Rosa (Schneider) Laturnus, Christina (Aberle) Long, Elaine Marie (Volk) Selensky, Marlene Ann (Kraft) Smith, Cora (Wolf) Tschaekofske, interview; Deloris (Boschee) Zimmerman, interview.

[28] Interview with Magdalena (Kuntz) Degenstein.

[29] Interview with Florence Engelhart.

Image 4.4. Young Gilbert Ost feeding the turkeys, *circa* late 1930s, showing how daily animal care taught German-Russian children responsibility, self-reliance, and pride in contributing to family life. *Photograph from the Gilbert Ost Collection, GRHC.*

ring cedar railroad ties for their intense heat. Farm labor laws at the time did not prohibit children from working alongside their families, and for Balzer, this work was simply part of growing up. He learned his responsibilities by watching and working alongside his father, gradually taking on more of the labor himself. Looking back, he insists they worked constantly—long days of hauling, chopping, and preparing fuel, ensuring the family had what it needed to endure the harsh North Dakota winters.[30]

Like Balzer, Joseph Michael Meier of Tramping Lake, Saskatchewan was responsible for assisting his father in all aspects of farming—plowing, seeding, repairing, and harvesting. There was always something to do on the farm, and it was typically a slow process. Joseph's father, for whom he is named after, kept Joseph and the older Meier boys out of school quite often to help on the farm. Joseph reflects upon that as quite beneficial, since once they finished school, the Meier boys stayed on the farm. These boys were receiving a practical education, while learning their gender role in the family, as well as society.[31]

[30] Interview with Balzer Kurtz.

[31] Interview with Joseph Michael Meier.

Then again, many narrators of the Dakota Memories maintain that during their childhood, they labored in the fields alongside their parents, regardless of gender.[32] While families valued traditional gender roles, the realities of farm life often made them impossible to maintain. For most German-Russian families, survival required flexibility, even if it went against expectations. Esther (Haerter) Drefs remembers that, despite the norm of keeping girls indoors for domestic tasks, she and her sisters worked alongside her father in the field at their home in Venstil, Montana. This obligation was a reality her father, Gottlieb, strongly disliked. He openly voiced his frustration, insisting that girls should focus on household work. However, necessity overruled tradition. With a sick wife and no other options, he had little choice but to allow his daughters to help, even if he refused to acknowledge it. Esther recalls him grumbling about it, and often trying to handle matters himself. And, yet, she and her sisters worked with the horses, tended crops, and handled tasks typically assigned to men, ensuring that the family could keep up with the unending demands of agricultural life.[33]

Florence (Mitzel) Engelhart, from Strasburg, North Dakota, shares her memories of shocking and haying during her oral history. Shocking involved gathering and stacking bundles of grain into upright piles, called shocks, to dry before threshing. This labor-intensive process helped protect the grain from moisture and made it easier to harvest. Haying, on the other hand, was the process of cutting, drying, and stacking hay to feed livestock during the winter months. Both tasks required strength and endurance, as they involve lifting heavy bundles, working in the heat, and keeping up with the relentless pace of farm life. Florence recalls spending long days in the fields performing these duties, work that was essential to both crop production and livestock survival. She brings to light a personal memory about getting stuck on top of a haystack, while setting the haystacks. One time, her father decided to leave her on top of the haystack while he went to get more hay from the field. Shortly after her father left, the weather shifted to thunder and lightning. Florence, who was deathly afraid of heights, was too scared to climb down. Eventually, her father returned, calmed her down, and carried her to safety.[34]

---

[32] Interviews with Arthur R. Bender, Polly Berger, Alice Buck, Edna Buck, LeRoy Buck, Stanley Graven Dobler, Esther (Haerter) Drefs, Florence (Mitzel) Engelhart, Clara Ferner, Henrietta Fiechtner, Ruth (Dockter) Fischer, Magdalena Hauck, Eileen Ketterling, Irene Schauer, Alma (Janke) Schott, Smith; William "Bill" Wald, Mary (Mitzel) Weber.

33 Interview with Drefs.

34 Interview with Florence Engelhart.

Typically, men and boys operated the horse-powered or machine-powered farm equipment; however, if the occasion called for it, women and girls operated farm machinery, too. Clara Ferner, for example, remembers learning how to drive the tractor and binder before turning eleven. She is certain of this timeline, as her family moved from their home in Kronau, Saskatchewan, when she was eleven, and she had already been using the equipment prior to their relocation.[35] Ruth (Dockter) Fisher, from Streeter, North Dakota, also recalls operating machinery. During threshing season, she was responsible for hauling the wheat into town, which was approximately seven miles from their farmstead. She would then load up the wagon with coal and head home. The experiences of Clara and Ruth highlight the flexibility of farm labor, where necessity often outweighed tradition, and young girls took on roles typically reserved for men to keep their families and farms running.[36]

Furthermore, children between the ages of six and twelve helped with all household responsibilities, but the division of labor varied depending on family size, location, and necessity. While traditional gender roles often shaped tasks, the demands of farm life frequently overruled expectations. On most family farms, children made their beds, swept the floors, washed the dishes, dusted the furniture, and cared for younger siblings. In some families, boys and girls followed expected roles, but in others, work was dictated by immediate need. Herbert Herman, an only child, recalls how his chores in Gackle, North Dakota, defied cultural expectations with his mother often reminding him, "Herbert, you got to be a girl and a boy!" Unlike boys in larger families who primarily worked outdoors, Herbert's responsibilities extended across household and farm work—spading the garden, fetching water, running errands, collecting coal, raising the chickens, and dusting the house. Like many German-Russian children, he did what was necessary, not what was traditionally expected, reinforcing the community's deeply rooted commitment to resilience and survival.[37]

Some narrators of the Dakota Memories recall that, while cooking and baking were traditionally considered domestic tasks, men and boys in their families often took part out of necessity.[38] Lawrence Johner, who lived in Kerrobert, Saskatchewan, fondly remembers spending time in the kitchen with his mother. He often helped her bake bread, sometimes making more than twelve

35 Interview with Clara Ferner.
36 Interview with Ruth Fischer.
37 Interview with Herbert Herman.
38 Interviews with Reinhold Bendewald, Lawrence Johner, and Peter Nagel.

loaves at once—an essential task to keep the household fed.[39] Ruth (Dockter) Fischer recalls that in the Dockter family, cooking was not just women's responsibility. With so much work to be done on the farm, her mother made sure all the men knew how to cook, ensuring that meals could still be prepared when other chores pull the women away. These memories reflect the practicality of farm life, where survival often depended on shared labor rather than rigid gender roles. [40]

The German-Russian narratives also document the blurring of gender lines for butchering and sausage making.[41] Marilyn Jeanette (Sauer) Bauman, from Java, South Dakota, insists her father was the best sausage maker in the area. She enjoyed helping him with the process, cutting the meat and the stuffing the casings. She, however, did not want any part of the slaughtering process; she could not even stomach being close by as the cows and hogs met their demise.[42] LeRoy Buck also remembers assisting with butchering and sausage making during his childhood in Streeter, North Dakota. When the Bucks had their sausage-making parties, LeRoy was responsible for washing the intestines. He had to pour hot water through the casings and scrub them.[43] Marilyn and LeRoy's memories highlight how butchering and sausage-making, though traditionally considered men's work, sometimes required the contributions of the entire family, with each person taking on tasks they could manage, even if some, like Marilyn, preferred to keep their distance from the more gruesome aspects.[44]

Looking back on their childhoods, narrators of the Dakota Memories appreciate the opportunities chores provided them. Many initially describe chores as grueling, time-consuming, and relentless, particularly during times of economic hardship and environmental struggle. Pauline "Polly" (Wock) Berger of Stark County, North Dakota, experienced firsthand the challenges of farming in extreme drought conditions, when families relied heavily on their gardens to feed themselves. Despite the difficulties, Polly treasured the time she spent gardening for her family. She enjoyed the fresh air, the feel of warm earth beneath her hands, and the satisfaction of watching plants grow. With little leisure time,

39 Interviews with Reinhold Bendewald, Lawrence Johner, and Peter Nagel.

40 Interview with Ruth Fischer.

41 Interviews with Smith, LeRoy Buck, Dewald, August Heyne, Klundt, Koepplin, Konrad, Krismer, Kurtz, Magdalena Miller, and LeRoy Presler.

42 Interview with Marilyn Bauman.

43 Interview with LeRoy Buck.

44 Interview with Marilyn Bauman.

she found joy in tasks like picking potatoes under the warm sun, harvesting watermelon with her siblings, and eagerly participating in garden contests to see who could grow the biggest radish or the crispest lettuce. These experiences nurtured a deep appreciation for gardening that lasted a lifetime. However, not all tasks carried the same fondness; her recounted her memories of housework, such as making beds, scrubbing floors, and washing dishes, lack the warmth and nostalgia of her time spent in the garden.[45]

Working with the animals was one aspect narrators consider especially memorable. Alice Ruth (Miller) Buck, from Streeter, North Dakota, enjoyed "slopping the pigs" and feeding the chickens, ducks, and geese. She found particular delight in feeding the pigs because of the messy, gooey foods they were given—grain and milk with a "cottage-cheesy" topping that pigs devoured eagerly. Recalling this moment with sensual delight, she describes the thick texture, the sour smell of fermenting milk, and the excited grunts of the pigs as they slurped up their meal. Alice's face glows as she reminisces about one of her favorite childhood memories.[46] For many children, tending to animals was more than just a responsibility—it was an experience that shaped their relationship with farm life.

From Morton County, North Dakota, Jacob A. Helbling, like Alice, has vivid memories of working with livestock. During his oral history session, Jacob reminisces about the envy he felt toward his older brother, who was old enough to ride the Shetland pony. The first time Jacob rode the pony, he fell off, but he enjoyed working with the horses so much that he and his brother even started breaking horses for other people to earn some extra money.[47]

For children raised on the plains, these moments—whether playful or challenging—were not just chores but defining experiences that deepened their connection to the land and animals they depended on.

Being able to work with the farm equipment was another memorable experience for German-Russian children. Although it was traditionally more common for boys to operate machinery, necessity sometimes required girls to take on these roles as well, as seen in the earlier account of Clara Ferner learning to drive a tractor at a young age. Still, for many boys, handling equipment was not only expected but also a source of pride. Edwin "Emil" Dewald from Napoleon,

---

[45] Interview with Polly Berger.

[46] Interview with Alice Ruth (Miller) Buck.

[47] Interview with Jacob A. Helbling.

Image 4.5. Elmer D. Sperling's uncle, Henry Sperling, working with horses as a young lad, symbolizing the trust, skill, and early independence expected of German-Russian children on the plains. *Photograph from the Elmer D. Sperling Collection, GRHC.*

North Dakota, loved operating the machinery and working with the tractor.[48] Simon Schwan, who is from Rugby, North Dakota, would wholeheartedly agree with Emil; he enjoyed growing up on a farm. He brags that by the age of eleven, he was riding a binder, and by the time he was twelve years old he was plowing with a two-bottom plow pulled by five horses. He was pleased and proud of the responsibilities he had as a young boy.[49]

Not all memorable experiences were pleasurable, as narrators of the Dakota Memories demonstrate. Jacob Helbling was an active participant on his family's farm, where he often spent time working with horses and operating

[48] Interview with Edwin "Emil" Dewald, interview.
[49] Interview with Simon Schwan.

horse-powered equipment, like a three-bottom plow, a header, and a binder. These tasks were part of his daily life, and his memories of handling the powerful horses and maneuvering the heavy machinery remain strikingly clear. Yet, while he looks back on endless chores with familiarity, not all of his experiences were without hardship. He recounts one particularly painful moment when something spooked his team of horses, causing them to bolt. In the chaos, they dragged the farm equipment over Jacob's foot, breaking it. With no medical attention, he had no choice but to endure the pain and allow it to heal on its own—an experience that reinforced just how dangerous farm work could be for a child. Through stories like his, Jacob highlights the reality that farm labor was not just demanding, but at times, unforgiving.[50]

Jacob's broken foot was just one example of how quickly accidents could happen on the farm. Many narrators recall moments when routine chores turned dangerous in an instant. Eileen M. (Burlack) Ketterling, from Burnstad, North Dakota, shares a frightening experience that happened while her family was helping a relative harvest wheat. She was driving the header-box; her mother, Fredericka (Ritter), was guiding a header pulled by six horses; and her sister, Mary, was stacking the wheat with the pitchfork. Everything was going as expected until, suddenly, Mary was gone. In an instant, she had slipped off the header-box—still holding the pitchfork. The moment was terrifying, but fortunately, she escaped without injury. Though unharmed, Mary was visibly shaken, a reminder of how quickly accidents could happen in the fields.[51]

The dangers of growing up on a farm were associated not only with the farm implements but also with the farm animals. Calvin K. Fercho, from Bismarck, North Dakota, shares his narrow escape from a dangerous encounter—not with a wild animal, but with his own pony. One time, while Calvin was out in the field rounding up the cattle, something spooked his pony, and it responded by kicking him in the face, knocking him unconscious. Though he doesn't remember the moment of impact, he recalls regaining consciousness and learning what had happened from his friend, who had witnessed the accident. It was fortunate his friend was there to ride back to the farm and get help. Calvin concludes his story by revealing the scar he still carries above his eye, a lasting reminder of the unpredictable risks of farm life.[52]

[50] Interview with Jacob Helbling.
[51] Interview with Eileen M. (Burlack) Ketterling.
[52] Interview with Calvin K. Fercho.

For children on the northern plains, as the Dakota Memories narrations reveal, chores did not disappear as they transitioned from childhood to adolescence (thirteen years of age until reaching adulthood, whether by leaving home or taking on greater responsibilities on the farm). According to narrators, both the difficulty of chores and parental expectations increased with age. Parents no longer expected their children only to do tasks like feeding cattle, cleaning barns, hauling water, gathering cow chips, and doing the laundry. Instead, as they got older they were expected to continue with those chores while taking on more significant responsibilities around the farmstead. For some, this meant preparing for eventual independence, while for others—especially sons who stayed to work the family farm—it was a transition into their future livelihood. Parents began teaching their children at a young age to manage the farm, ensuring they learned the necessary skills before adulthood. During their adolescent years, German-Russians recall gaining a deeper understanding of these future responsibilities as people of the plains.[53]

The Dakota Memories narrations reaffirm and expand the existing information on childhood chores and responsibilities in the American West. Children worked hard, and the work they did had purpose. Chores taught them responsibilities, traditional roles, gender roles, and character traits. The work allowed them to contribute to their families' finances, and as a result their families' survival, and gave them pleasurable (and traumatic) experiences to share.[54] The narrators reveal that Viviana Zelizer's "priceless child" concept was inapplicable for German-Russian families of the mid-twentieth century.[55] This does not mean German-Russians did not value or love their children. It was, in fact, quite the opposite. The first-generation Germans from Russia expressed their love and affection by instilling in their children a strong work ethic and providing them with the necessary skills to survive on the plains. Beyond simply learning chores, children developed the resourcefulness, resilience, and adaptability required to sustain a household in an unpredictable and often unforgiving environment. Managing livestock, preserving food, hauling water, and tending to fields were not just daily tasks—they were essential for enduring long winters, droughts, and economic uncertainty. These oral histories

---

[53] Interviews with Magdalena (Staudinger) Hauck, Louis Helfrich, Michael Jahner, and Edward Matthew Selensky.

[54] For more information, see West, *Growing up with the Country*; Riney-Kehrberg, *Childhood on the Farm*; Holt, *Children of the Western Plains.*

[55] Zelizer, *Pricing the Priceless Child*, 3.

also reveal that childhood was neither as rosy as West paints it, nor as grim as Hampsten portrays it. Rather German-Russian childhood was nuanced, closely resembling the descriptions of Holt and Riney-Kehrberg. The Dakota Memories project not only preserves these personal accounts but also enriches our understanding of how labor shaped identity, family dynamics, and cultural continuity among German-Russians. The narratives provide a vital perspective on how work functioned as both necessity and heritage, reinforcing that for these families, survival and tradition were deeply intertwined.

The selected memories of chores and responsibilities highlight the central role of labor in shaping the childhood experiences of the second- and third-generation Germans from Russia. While narrators often describe chores as exhausting and relentless their individual stories reveal a more precise reality. Children's work extended beyond simple necessity. Rather, it was a means of learning essential skills, contributing to the household, and strengthening family bonds. Boys and girls alike took on demanding tasks, from feeding livestock and hauling water to cooking, sewing, and operating farm machinery, though the division of labor was often dictated by tradition, family needs, or circumstance. As they grew older, responsibilities increased, preparing them for adult life on the plains, whether they remained on the farm or pursued other paths. Through these memories, narrators illustrate that childhood for German-Russians was neither defined solely by hardship nor idealized as carefree. Instead, it was shaped by resilience, adaptability, and a deep sense of duty—values that remained central to their collective identity.

For many narrators, looking back on their childhood work reveals a shift in perspective—from seeing chores as burdensome, to recognizing their lasting value. Hilda Melita (Lindemann) Backfish, of Golden Valley, North Dakota, believed she was overworked and underpaid as a child. However, as she reminisces about her chores and responsibilities, she acknowledges that, in hindsight, she has nothing to complain about. What once felt like endless labor became, in retrospect, a foundation for resilience and appreciation. Hilda concludes her story with a phrase that captures the German-Russian work ethic: *Arbeiten macht Lieben Süß*, or, "Work makes life sweet."[56]

[56] Interview with Hilda Melita (Lindemann) Backfish.

CHAPTER 5

Reading, Writing, and Arithmetic: Grade School

When Jacob A. Helbling of Morton County, North Dakota, was six years old, he embarked upon a new adventure: grade school. He remembers attending a country school, a one-room schoolhouse. It was small and crowded, with sixteen or seventeen students (from all eight grades) and one teacher. Thinking back on his childhood, Jacob describes the frigid winters in that one-room schoolhouse, Rural School #01 or Rural School #02. Jacob attended both schools at different stages in his childhood and cannot differentiate between his early memories of both places. Regardless, Jacob says it was very cold in the mornings. Teachers encouraged all children to gather around the potbelly coal stove upon arrival. Some of the boys would rest their shoes so close to the stove that smoke began to rise from their feet. Chuckling, as if he was reliving the experience, Jacob insists most mornings the room smelled like burning rubber. He remembers teachers reminding them to put distance between their shoes and the stove.[1]

Jacob, like other narrators of this project, has vivid memories of attending grade school on the northern plains.[2] They reminisce about a variety of topics, such as schoolhouses, teachers, curriculum, recess, noon hour, and attendance. Furthermore, Dakota Memories narrators acknowledge that schools were agents of acculturation; they insist schools promoted assimilation. In the United States, schools were supposed to Americanize children, while Canadian schools were supposed to Canadianize them. The reality, however, was more

[1] Interview with Jacob A. Helbling.
[2] Interview with Jacob Helbling.

complex. As narrators attest, these one-room country schools reinforced their shared memory and identity.

While there were restrictions set in place to transform the identity of immigrant children, such as the use of native language and limited unsupervised free time, schools in the predominately German-Russian rural communities brought immigrant children together in environments where they shared a common language, customs, and traditions. Rather than erasing their ethnic identity, these schools reinforced it, allowing traditions to persist and adapt within a familiar cultural setting.

Childhood historians, like Elliott West and Marilyn Irvin Holt, provide informative yet cursory analyses of grade school in their works.[3] Wayne Fuller and Pamela Riney-Kehrberg, however, provide more in-depth examinations.[4] In his article, "School District 37," Fuller recognizes the significant role schools played in rural communities. He asserts that these institutions were instrumental in developing community and fostering regional and ethnic identity. Schools were, in Fuller's words, "a community's cohesive force."[5] In her work, Riney-Kehrberg also describes the role this institution played in the lives of farm children, finding that schools were one of the most significant influences on children, second only to chores. She acknowledges that "in terms of devotion to school, education could be an element of little importance to a youngster, or of the utmost importance."[6]

In the mid-1800s and mid-1910s, territories, states, and provinces had compulsory school attendance. Law made this public institution central in the lives of children. Nonetheless, "schools, the school day, and the academic year," as Riney-Kehrberg points out, "varied from location to location."[7] The compulsory laws established mandatory attendance, but they did not regulate much else. Farmers became responsible for establishing their own schools, including the location, building administration, teacher, policies, and funding. Fuller writes, "the law gave them responsibility of electing their own school board members,

[3] For more information, see West, *Growing up with the Country*; Holt, *Children of the Western Plains.*

[4] For more information, see Riney-Kehrberg, *Childhood on the Farm*; Wayne E. Fuller, *The Old Country School: The Story of Rural Education in the Middle West* (Chicago: University of Chicago Press, 1982); Fuller, "School District 37: Prairie Community," *The Western Historical Quarterly* (October 1981): 419–32.

[5] Fuller, "School District 37," 432.

[6] Riney-Kehrberg, *Childhood on the Farm*, 62.

[7] Riney-Kehrberg, *Childhood on the Farm*, 62.

Image 5.1. Spruce Hill #62, *circa* 1940s, one of many rural one-room schools that served as educational and cultural hubs for German-Russian children, fostering community pride and adaptation.
Photograph from the Marlene Ann (Kraft) Smith Collection, GRHC.

determining how many months—beyond the required three—they would hold school each year and the amount of taxes they would raise annually to pay for the education of their children."[8] In essence, schools were internal infrastructures, and as a result their oversight varied from community to community. Yet, as Riney-Kehrberg notes, there existed, "enough similar characteristics that good generalizations can be made."[9] For instance, similarities existed between the schoolhouses, teachers, curriculum, recesses, attendance, and value. Despite regional differences, the structure and purpose of rural schools remained

[8] Fuller, "School District 37," 423.
[9] Riney-Kehrberg, *Childhood on the Farm*, 62.

consistent—serving as both educational institutions and community hubs that reflected the values and needs of the families they served.

One-room country schoolhouses were simple, practical structures that were a common sight throughout the West. Typically, they featured rows of windows on one or two walls for natural light, with blackboards positioned at the front and back of the room.[10] Most had three windows along the north and south sides, and students sat in sturdy, patented desk that were considered a prized feature [11] While some schools only had a handful of students, others accommodated fifty or more, reflecting the variation in rural school populations.[12]

The quality of education ranged widely, as there were no uniform standards across rural schools. Each community set its own expectations for curriculum, teacher qualifications, and instructional methods, leading to significant differences in what students learned and how they were taught. Teachers, who were usually young and inexperienced, taught three main subjects: reading, writing, and arithmetic. They approached other subjects like history, geography, and grammar through lessons on one of the main subjects. Country schoolteachers usually had no more than a high school education and a teaching certificate. Furthermore, "teachers, who were often quite young and minimally trained," Riney-Kehrberg writes, were "unprepared to handle many of the discipline problems they faced in the classrooms."[13] And, as a result, maintaining order with students spanning eight grades in one classroom tended to monopolize the teachers' time throughout the school year.[14]

Nonetheless, it was through the instruction of the three Rs (reading, [w]riting, and [a]rithmetic) that Americanization or Canadianization took place.[15] In *The Minds of the West*, historian Jon Gjerde explores the role of schools in bringing immigrant children into mainstream society. He argues, "[T]he institution most instrumental in bringing about such [cultural] diffusion was the public school."[16] Then again, as Gjerde points out, "the influence of public school grew only to the degree that immigrant children attended it."[17] Once

---

[10] Riney-Kehrberg, *Childhood on the Farm*, 62.

[11] Fuller, "School District 37," 429.

[12] Riney-Kehrberg, *Childhood on the Farm,* 62.

[13] Riney-Kehrberg, *Childhood on the Farm*, 77.

[14] Riney-Kehrberg, *Childhood on the Farm*, 63–77.

[15] Jane Marie Pederson, *Between Memory and Reality: Family and Community in Rural Wisconsin, 1870–1970* (Madison, WI: The University of Wisconsin Press, 1992), 132.

[16] Gjerde, *The Minds of the West,* 233.

[17] Gjerde, *The Minds of the West*, 233.

territories, states, and provinces passed compulsory education laws, contention arose over the issues of assimilation and acculturation in public schools. Gjerde states that "rarely were rural ethnic communities so isolated and so consciously separate that they could avoid the influences of a quickening cultural evolution in the late nineteenth century."[18]

Immigrant families feared that American or Canadian schools, with their secular curriculum and broader societal influences, might weaken their children's connection to faith, tradition, and family values. Public education emphasized knowledge for its own sake rather than reinforcing moral or religious teachings, which many immigrants believed were central to raising responsible, community-oriented individuals.[19] While education itself was widely valued, immigrant parents—especially those of German heritage in this study—worried that schooling would lead their children away from traditional beliefs and encourage assimilation at the expense of cultural identity. These concerns extended beyond the classroom, as public schools often prepared students for careers that required them to leave home, undermining the agricultural lifestyle many parents had intended for them.[20] For rural families, particularly those tied to farming communities, this potential loss of both cultural heritage and future labor made public education a complex and, at times, unwelcome institution.

Although families feared the public school system, compulsory attendance laws required children to enroll, making public education unavoidable for most immigrant families. In response, some communities, particularly Catholic ones, established parochial schools that integrated religious instruction with academics, while public schools, including the small one-room schoolhouses common in rural areas, followed a more secular curriculum. Protestant families were more likely to send their children to public schools, where the emphasis was on civic responsibility, nationalism, and assimilation rather than ethnic or religious traditions. These schools, regardless of size, played a role in cultural integration, often requiring students to speak only English and discouraging practices tied to their heritage.

As Gjerde discusses, some families and communities sought alternatives to the public school system by establishing parochial schools, though this was more common among Catholic immigrants than Protestants. Most European immigrant children attended public schools, where they encountered firsthand

[18] Gjerde, *The Minds of the West*, 230.
[19] Gjerde, *The Minds of the West*, 275.
[20] Gjerde, *The Minds of the West*, 181.

the secular agenda of public education. Teachers acted as conduits of nonsectarian and national ideals, reinforcing mainstream cultural norms and expectations. For example, students were often prohibited from speaking their native language on school grounds, further accelerating their linguistic assimilation. The broader national agenda of uniformity was evident in public education policies, with teachers positioned as intermediaries between rural ethnic communities and the expectations of mainstream society. Through policies like English-only instruction, schools prepared children for participation in modern civic life, but in doing so, they also eroded aspects of their families' linguistic and cultural traditions, reinforcing immigrant concerns about identity loss.[21]

Neither historians of childhood (such as Riney-Kehrberg) nor historians of ethnicity and acculturation (such as Gjerde) address German-Russians outright in their studies, yet their findings are applicable to the history of this ethnic group.[22] Unlike their grandparents and parents, the second- and third-generation Germans from Russia—the Dakota Memories narrators—experienced country-school education shaped by national education policies. At some point in their oral histories, narrators reminisce about attending schools described as simple structures, receiving instruction from young and inexperienced teachers, playing games like Fox and Geese (a popular winter game where children stamped a wheel-like pattern into the snow and chased each other along its paths) during noon-recess hours, and learning English in school. They reflect on the efforts of teachers and school administrators to instill national ideals and their own personal resistance or adaptation to these influences. Their grade-school memories include a myriad of stories about the schoolhouses, teachers, curriculum, recess and noon hours, attendance, and the broader significance of education shaping their identities.

The descriptions of these traditional one-room schoolhouses by Dakota Memories narrators mirror the descriptions of Riney-Kehrberg and Fuller, tending to focus on the simplicity of the schoolhouse structure.[23] Clarence Bau-

---

[21] Gjerde, *The Minds of the West*, 181.

[22] For more information, see Riney-Kehrberg, *Childhood on the Farm;* Gjerde, *The Minds of the West.*

[23] For more information, see Fuller, *The Old Country School*; Fuller, "School District 37"; Riney-Kehrberg, *Childhood on the Farm.* Dakota Memories narrators, in fact, provide more detail in describing the other buildings, like outhouses, barns, and sheds, than they do for the schoolhouse itself. See interviews with Clarence Bauman, Alice (Woehl) Buerkley, Edna Lu Ella (Weisz) Buck, Ben Ferner, Marjorie (Miller) Flegel, Robert R. Gab, Magdalena (Staudinger) Hauck, Wilma (Beck) Haupt, Jacob A. Helbling, Gilbert Ost, Emil Roesler, Herbert Thurn, and Rueben Richard Wolf.

man, from Java, South Dakota, for example has fond memories of attending a one-room country school and describes his one-room schoolhouse as being small. He adds little more detail, estimating that the building was approximately sixteen feet wide and twenty-four feet long. In the entryway, he notes, there was a bin for coal to be used in the potbelly stove that was located at the rear of the schoolhouse. The desks in the room varied in size—smaller desks for younger children and bigger desks for older children. Clarence remembers that the "school marm" had her desk positioned next to the blackboard in the front of the classroom.[24]

Since the one-room schoolhouses were located in rural regions, children had to walk long distances to and from school. Dakota Memories narrators have memories of walking one, two, three, and even four miles. For the most part, narrators enjoyed their walks across the prairie, free from adult supervision. Then again, the northern plains environment sometimes hindered their fun.[25] Herbert Thurn, who hails from Venturia, North Dakota, remembers walking two to three miles when the weather permitted. He was the youngest of five children, which was a relatively small German-Russian family. On snowy days his father used to take the older Thurn children to school in a makeshift enclosed buggy. This practice ended before Herbert was old enough to attend school, as his father died unexpectedly in 1932. Thus, Herbert has no memories of riding to school with his father. He remembers, however, his older siblings using their father's invention to take him to school quite often. As soon as they arrived at the school, Herbert says, they unhitched the horses and fed them hay that they hauled from home, then scurried off to the schoolhouse in time for the day to begin.[26]

Hilda Melita (Lindemann) Backfish, from Golden Valley, North Dakota, also remembers her commute to and from school. The first country school she attended was located two miles west of her parents' farm. When that school closed, she attended one located two miles to the east. The Lindemann children usually walked to school, even in near blizzard conditions. Hilda points out they sometimes made a human train to get home. The oldest boy, William John,

---

[24] Interview with Clarence Bauman.

[25] Interviews with Hilda Melita (Lindemann) Backfish, Clarence Bauman, Buerkley, Edna Buck, Ben Ferner, Flegel, Jacob Helbling, Louis Helfrich, Elise (Just) Johnson, Victor Knell, Ost, Tillie (Deisz) Rath, Shirley Ann (Weigum) Stern, Herbert Thurn, and Cora (Wolff) Tschaekofske.

[26] Interview with Herbert Thurn.

Image 5.2. Students from the Alamo school hitching a ride home, date unknown, capturing the communal spirit, resourcefulness, and shared joy of rural school life on the northern plains. *Photograph from the Clarence and Marilyn Jeanette (Sauer) Bauman Collection, GRHC.*

was the "engine car" leading the rest of the children. He would carry his lunch box on his shoulder and lead the way home. Then, the youngest boy, Rudolph Frederich, was next, leaning on his brother and burying his face in Rudolph's coat. Then, Hilda and her sister, Leona Amelia, lined up behind Rudolph, forming the end of the train. They also buried their faces in the person in front of them to shield them from the inclement weather. The journey to school, as the Lindemann children experienced firsthand, was an adventure in itself.[27]

The arctic weather, as Dakota Memories narrators attest, made these little one-room country schoolhouses quite cold. School buildings had inadequate heating systems; as a result, people remember being cold throughout the school day. Typically, either potbelly stoves or furnaces heated the buildings, yet they rarely operated around the clock. Thus, schools tended to be cold first thing in

[27] Interview with Backfish.

the morning, especially Monday mornings.[28] Students, like Herbert Thurn, for instance, remember their teachers going in early to start the stove or furnace, trying to warm up the building before students arrived. Herbert also recalls his teacher banking the furnace with coal before leaving in the evenings in an attempt to keep the schoolhouse warm for as long as possible. The intent was to make the building less cold in the morning.[29]

The descriptions of one-room country schoolhouses by German-Russian descendants are much the same as the images recorded in the childhood historiography of the American West. The schoolhouses were rural, small, and inadequate. Yet, the similarities end there. The new German-Russian story deviates with descriptions of class size and ethnic make-up. According to Dakota Memories narrators, their class sizes were smaller, ranging from ten to thirty students, and they had a more homogenized group of mostly German-Russian students.[30]

Growing up on a farm near the small town of Fredonia, North Dakota, during the 1930s, Alice (Woehl) Buerkley attended a distinctly German-Russian country school. Alice insists "everybody was German" at her country school, including several of her teachers. Alice remembers about fifteen students, at different levels, attending her one-room schoolhouse. She also says that although most students and teachers were from German-Russian backgrounds, English was the only language permitted on school grounds. Alice remembers that her teachers (especially the ones with German backgrounds) allowed them to speak German, as long as they attempted to learn English during their lessons.[31] While most Dakota Memories narrators do not recall such leniency from their teachers, others do recall defying school rules by speaking German.[32]

Dakota Memories narrators describe their teachers with specificity, exposing similarities between teachers on the northern plains and teachers in the Midwest. Their narrations reveal that the image of schoolteachers provided by historians like Riney-Kehrberg and Fuller is applicable to schoolteachers on the plains. Schoolteachers in the German-Russian communities, as in most rural

[28] Interviews with Clarence Bauman, Edna Buck, Gab, Elsie Johnson, Roesler, Stern, and Herbert Thurn.

[29] Interview with Herbert Thurn.

[30] Interviews with Buerkley, Edna Buck, Ben Ferner, Flegel, Gab, Haupt, Elsie Johnson, Knell, Ost, Tillie Rath, Roesler, Tschaekofske, and Rueben Richard Wolf.

[31] Interview with Buerkley.

[32] Interviews with Backfish, Clarence Bauman, Buerkley, Ben Ferner, Jacob Helbling, Joseph Michael Meier, Ost, interview; Herbert Thurn, and Wolf.

communities, were young, inexperienced women. According to Riney-Kehrberg and Fuller, school boards thought it was ideal to bring in a teacher from the East. They believed teachers from urban centers in the East would bring worldly knowledge and sound morals to their country schools.[33] The reality was that school boards resorted to hiring primarily local girls (or local boys on occasion), although that was not always feasible. Memoirist Dolly Holliday Clark is a perfect example of the difficulties that resulted from hiring girls who were unfamiliar with the region and its inhabitants. Clark was a seventeen-year-old teacher from Indiana who in 1919 came to North Dakota to teach in a German-Russian community near Mandan. Throughout her memoir, Clark reminisces about the poor quality of education due to the lack of community support and funding, rather than her inexperience, and her tense relationships with members of an ethnic community, which led to her decision to leave after only one year.[34]

The Dakota Memories narrations reveal mixed feelings regarding the quality of education youngsters received from rural educators. Some narrators reminisce about their lack of a quality education, claiming their teachers were inexperienced and ill prepared. Others recognize and admire their teachers for succeeding in difficult conditions.[35] Growing up in the 1930s near Glen Ullin, North Dakota, Louis Helfrich has memories of good teachers and bad teachers. Louis remembers he had exceptional teachers when he attended a country school in Morton County. He says they were patient, and he credits them for teaching him to read. Then, his family moved to Grant County, where his education deteriorated. The teachers never stayed around longer than a month or two, making it nearly impossible for children to retain lessons. Louis blames the economy of the 1930s for the uneven education provided in Grant County. According to Louis, the school boards paid teachers in "warrants"—a form of government-issued promissory notes that functioned as a deferred payment—but there was no money to back them up. Thus, teachers "fled the region" as soon as they realized they were working for free. Louis cannot remember exactly how

---

[33] For more information, review Riney-Kehrberg, *Childhood on the Farm*; Fuller, *The Old Country School.*

[34] Dolly Holliday Clark, "Memoir of a Country Schoolteacher: Dolly Holiday Meets the Ethnic West, 1919–1920," *North Dakota History: Journal of the Northern Plains* 59 (Winter 1992): 30–45.

[35] Interviews with Backfish, Clarence Bauman, Buerkley, Edna Buck, Ben Ferner, Gab, Margaret Gross, Haupt, Magdalena Hauck, Jacob Helbling, Louis Helfrich, Donald T. Huber, Elise Johnson, Knell, Ost, Tillie Rath, Roesler, Stern, Herbert Thurn, Wolf.

many teachers he had while attending school in this county, but he insists it was common to have multiple teachers in any given year.[36]

At the same time, in Dickey County, Marjorie (Miller) Flegel was having similar experiences in her one-room country school. She remembers having some good teachers, and some not-so-good teachers. A number of teachers, she recalls, enjoyed having older students in the same room as younger students. These teachers divided their classes into groups, so the younger students could learn by either listening to or interacting with the older students. Then again, Marjorie points out, not all of her teachers appreciated the opportunities presented by their rural setting. She remembers teachers simply biding their time until the end of the school year. There was one teacher, she remembers, who simply disappeared for days at a time, without any word to the school board, parents, or students. Eventually, the school board dismissed her, and appointed a substitute to finish the year.[37]

According to historians such as Riney-Kehrberg and Fuller, some teachers in one-room country schools came from urban centers in the East and boarded with local farm families near the schoolhouses.[38] This arrangement often presented cultural and social challenges, particularly in tight-knit, homogeneous communities like those of the German-Russians'. Dolly Holliday Clark, for instance, describes her experience boarding with the Mueller family in her memoir, calling it "a rude awakening." Unfamiliar with the customs and lifestyle of her host family, she initially struggled to connect with the children—who were also her students—and never established more than a formal relationship with their parents. After several months, she moved in with another local family, who happened to be relatives of the Muellers, but by the end of the school year she decided to leave the community altogether.[39] Teachers, like Clark, faced a unique challenge: they were outsiders, tasked with assimilating their students into mainstream culture, yet they lived and worked within insular ethnic communities that often remained resistant to outside influence.

Boarding with a nearby farm family was not just a necessity for country schoolteachers during the winter months—it was often the only option. Even local teachers needed to establish a residency close to the schoolhouse. Some

---

[36] Interview with Helfrich.

[37] Interview with Flegel.

[38] For more information, consult Riney-Kehrberg, *Childhood on the Farm*; Fuller, *The Old Country School.*

[39] Holliday Clark, "Memoir of a Country Schoolteacher," 30–45.

narrators remember teachers staying with neighbors or friends, while other narrators recall teachers boarding with their families. Regardless, from the children's perspective, these boarding experiences were typically positive. Having a teacher reside with them meant opportunities to establish solid relationships, to receive extra help on homework, and for regular transportation to and from school.[40] Alice (Woehl) Buerkley remembers forming strong relationships with the teachers who boarded at the Woehl farmstead. Her teachers, Alice says, always spoiled them. One of her teachers always held her younger brother, Arthur, during reading lessons, helping him to stay focused on the material.[41]

Marjorie Flegel also remembers teachers boarding at her parents' place while she was going to school. According to her, "it was great." Marjorie enjoyed it, because her close proximity to the teacher provided an opportunity for more socializing at home, and she received extra assistance with her assignments. In the Miller home, boarders usually had their own bedroom, but partook in the family meals and shared the family living space. Marjorie remembers going to school and returning home with the teachers most days, whether it was on foot, horse, or car. She points out that when the snow was deep her father typically took care of transporting his children and the teacher to and from school.[42]

For the most part, Dakota Memories narrators recall their teachers being local girls (and occasionally local boys). They remember having a good deal of teachers from within their own ethnic group.[43] Thus, Germans from Russia were not resisting their nation's compulsory education and were, in fact, buying into the American or Canadian school system. Gilbert Ost, from Beulah, North Dakota, and Clarence Bauman, for example, remember having teachers of German-Russian heritage. They insist teachers who were of German-Russian ancestry were easier to get along with, as they could relate to each other. Gilbert recounts his memories of Ms. Olson, the only non-German-Russian teacher he remembers. According to Gilbert, she was the toughest and meanest teacher from his childhood. She was the only teacher Gilbert disliked. Part of his contempt for Ms. Olson stems from her discipline policies for using German language on school grounds. She made children write, "I will not speak German," more than once on the blackboard, as a punishment for breaking the

[40] Interviews with Buerkley, Edna Buck, Flegel, and Stern.

[41] Interview with Buerkley.

[42] Interview with Flegel.

[43] Interviews with Clarence Bauman, Buerkley, Jacob Helbling, Ost, Tillie Rath, Roesler, and Herbert Thurn.

school rules. Looking back as an adult, Gilbert interprets his frustration as more than just a child's resentment toward a strict teacher. He now sees it as part of a larger struggle—the tension between school policies and the desire to hold onto his cultural identity. While he may not have fully understood this dynamic as a child, his memory of those punishments still carries the weight of that conflict.[44]

Before teachers in these ethnic enclaves could begin instructing children in the basics of reading, writing, and arithmetic, most needed to introduce their students to English. It was essential that teachers instruct their students in this language, especially if they were unfamiliar with the immigrant group's native dialect. Furthermore, introducing children to English was the first step in the assimilation or acculturation agenda of the public school.[45] Thus, learning English (or sometimes French in Canada) was an essential element of the Americanization or Canadianization agendas of national governments.[46] According to Gjerde, "English was typically the explicit language of instruction" even in ethnic communities where English was not the predominate language.[47] German-Russians of the second and third generation substantiate this claim. Most narrators recall starting school without ever having spoken a word of English, as German was typically the predominant language in German-Russian communities. Even Dolly Holliday Clark notes that her first-grade pupils in Mandan, North Dakota, only spoke German upon entering school.[48]

Countless Dakota Memories narrators attest to the difficulties of starting school without knowing the English or French language.[49] Joseph Michael Meier of Tramping Lake, Saskatchewan, remembers the difficulties he faced by starting school unable to speak English.[50] Since school boards and teachers only permitted English on school grounds, this transition was especially difficult for immigrant children growing up in homogenous ethnic communities. As Joe remembers, mastering this new language was incredibly hard, since his teachers were not familiar with German, and they prohibited the use of Ger-

---

[44] Interview with Ost.

[45] For more information, see Gjerde, *The Minds of the West.*

[46] Pederson, *Between Memory and Reality,* 132.

[47] Gjerde, *The Minds of the West*, 223.

[48] Holliday Clark, "Memoir of a Country Schoolteacher," 34.

[49] Interviews with Backfish, Clarence Bauman, Buerkley, Edna Buck, Ben Ferner, Flegel, Jacob Helbling, Meier, Ost, Herbert Thurn, and Wolf.

[50] Interview with Meier.

man in school. This circumstance made communicating difficult for quite some time, Joe recalls.[51]

Occasionally narrators reveal a covert resistance to this so-called forced transition to English. For instance, they spoke German during recess or noon hour, and when they played games or took part in social activities.[52] As a result, "school classrooms frequently became sites of cultural conflict between pupils and native-born teachers," says historian Steven Mintz. Mintz continues, these teachers tended to attach "little value to cultural diversity and sought to transform immigrant children into 'little citizens' by forbidding them to speak their native tongue."[53] German-Russian children felt more comfortable conversing in German, so they refused to abandon their language outside of the classroom.

For some, however, avoiding English was not an option. In his oral history Ben Ferner, who was born in Katherienthal, Saskatchewan, recalls how teachers enforced English-only policies not only during lessons but even during recess and school programs. Since German was the only language he knew, following this rule proved nearly impossible. He remembers one instance when his teacher caught him speaking German while playing Xs and Os on the blackboard, earning him "the strap" for the first time.[54] Despite these experiences, Ben and other narrators acknowledge that their teachers' persistence ultimately helped them become bilingual. In the end, most express no resentment toward their teachers for this so-called forced transition.[55]

As noted earlier, rural schoolteachers primarily focused on the basics—reading, writing, and arithmetic—though other subjects, such as civics, nature, and agriculture, were sometimes included varying by state or province.[56] Regardless of the curriculum, lessons introduced immigrant children not only to academic subjects but also to the culture of their new *heimat* (homeland).[57] Dakota Memories narrators frequently recall the basic subjects as the founda-

---

[51] Interview with Meier; Jessica Clark and Thomas D. Isern, "Germans from Russia: An Oral History," 79.

[52] Interviews with Backfish, Clarence Bauman, Buerkley, Edna Buck, Ben Ferner, Flegel, Jacob. Helbling, Meier, Ost, Herbert Thurn, and Wolf.

[53] Mintz, *Huck's Raft*, 204.

[54] Interview with Ben Ferner.

[55] Clark and Isern, "Germans from Russia," 79.

[56] For more information, see Riney-Kehrberg, *Childhood on the Farm*; Fuller, *The Old Country School.*

[57] Mintz, *Huck's Raft*, 208.

Image 5.3. Students of Faika School #3, with Alma (Janke) Schott (a Dakota Memories interviewee) on the far right, back row, *circa* 1930s, illustrating the community pride and friendships forged in rural school life. *Photograph from the Alma (Janke) Schott Collection, GRHC.*

tion of their education, but some remember additional subjects.[58] Cora (Wolff) Tschaekofske, who attended grade school in Golden Valley, North Dakota, in the 1930s, recalls learning spelling, geography, and health alongside the core subjects. While she, like many others, saw education as a means of learning practical skills, she also recognized its broader influence in shaping her understanding of the world—particularly through history, her favorite subject.

Like many rural students, Emil Roesler, from Lehr, North Dakota, recalls a school curriculum that was both structured and repetitive, reinforcing key subjects through memorization and recitation. He recalls learning history and civics—his favorite subjects—as well as geography, arithmetic, spelling, and

[58] Interviews with Backfish, Clarence Bauman, Edna Buck, Ben Ferner, Flegel, Gab, Huber, Victor Knell, Ost, Tillie Rath, Roesler, Stern, Herbert Thurn, Cora (Wolff) Tschaekofske, and Wolf.

hygiene, which was typically taught as part of health education. However, he never enjoyed geography, and spelling was not his strongest subject. The method of teaching across all subjects, as Emil's memories demonstrate, relied heavily on rote memorization. He provides an example from his spelling lessons, describing how his teacher would stand in front of the classroom, call out a word aloud, and instruct students to write it down using the correct spelling. Then, students exchanged papers to correct one another's work. Emil remembers the outcome all too well, "I always had some wrong." Despite the challenges, he and his classmates accepted these teaching methods as a normal part of their education, shaping not only their academic knowledge but also their discipline and perseverance.[59]

Although narrators mainly recall learning the subjects of reading, writing, and arithmetic, the seventh and eighth grade exams demonstrate that teachers taught other subjects, as well. Edna Lu Ella (Weisz) Buck, who grew up near Lehr, North Dakota, in the 1920s and 1930s, remembers taking three written exams to pass the seventh grade and written exams in hygiene, agriculture, math, and history to pass the eighth grade.[60] Emil Roesler, also from Lehr, recalls that one of his exams focused on geography.[61] Elsie (Just) Johnson, originally from Wisconsin, points out that these exams were rites of passage for grade school students. They had to pass the exams in order to complete the eighth grade. If they failed, they were doomed to repeat grade school the following year. Elsie recalls there were always a few stragglers taking the eighth grade for an additional year or two (or more).[62]

Country schools were more than simple institutions designed for education and acculturation or assimilation. They were also institutions that fostered community development and socialization. Furthermore, Fuller writes, these one-room schoolhouses created homogeneity amongst rural farm children.[63] Dakota Memories narrators would agree with Fuller. Schools provided children with opportunities to leave the farm and make friends with other children from similar backgrounds. For German-Russian children this usually meant meeting others from within their ethnic group. While schoolteachers spent most of the day teaching curriculum designed for acculturation and assimilation, recess-

---

59 Interview with Roesler.
60 Interview with Edna Buck.
61 Interview with Roesler.
62 Interview with Elsie Johnson.
63 Fuller, "School District 37," 419, 424.

es and noon hour provided students free time for children's play. "Morning and afternoon recesses, as well as lunches," in the words of Riney-Kehrberg, "were the only regularly scheduled reliable time for play that farm children had," and, as attested by the Dakota Memories narrators, for forming close relationships."[64]

Noon hour was one of the main highlights of grade school, offering a full hour of unstructured playtime. Children often rushed through their meals to maximize time outdoors, eager to play games, explore, and socialize. As Riney-Kehrberg notes, "[I]t was the entertainment value of school that made attendance worthwhile."[65] Dakota Memories participants would agree—many insist noon hour was the best part of the school day, reminiscing in detail about everything from their lunch pails to their favorite outdoor games. For most, the meal itself was routine, a mix of home-packed staples reflecting their families' farm-based diets.[66] Victor Knell, from Hazen, North Dakota, attended Krem School #4 from 1947 to 1955. He remembers bringing homemade bread with sausage, jam, or cheese, sometimes a cookie or an apple. "It wasn't anything special," he recalls, as the real excitement came after eating.[67] Unsupervised time allowed children to engage in rowdy games like Fox and Geese, Anti-I-Over, and tag, climb trees, or explore the surrounding landscape. Older students—especially those in sixth or seventh grade or twelve or thirteen years old—became more aware of social dynamics, using this time to tease, flirt, or speculate about relationships. These stolen moments of independence, away from the watchful eyes of teachers, made noon hour truly unforgettable.

Farm children spent their free time—whether during recess or noon hour—playing games, socializing, and occasionally getting into mischief, both inside and outside the schoolhouse. These unstructured moments were not just about fun; they were an essential part of the school experience, shaping friendships and reinforcing a sense of belonging among young immigrant descendants. Unlike play at home, which was often solitary or limited to siblings, school provided a setting where children could interact with a larger peer group, learning

---

[64] Riney-Kehrberg, *Childhood on the Farm*, 82; Interviews with Backfish, Clarence Bauman, Buerkley, Edna Buck, Ben Ferner, Flegel, Gab, Jacob Helbling, Elsie Johnson, Knell, Tillie Rath, Roesler, Stern, Tschaekofske, and Wolf.

[65] Riney-Kehrberg, *Childhood on the Farm*, 82.

[66] Interviews with Backfish, Clarence Bauman, Buerkley, Edna Buck, Ben Ferner, Flegel, Gab, Wilma (Beck) Haupt, Jacob Helbling, Huber, Elsie Johnson, Knell, Ost, Tillie Rath, Roesler, Stern, Tschaekofske, and Wolf.

[67] Interview with Knell.

Image 5.4. Dakota Memories participant Marlene Ann (Kraft) Smith playing during recess with friends, *circa* 1946–1947, capturing the joy, social bonds, and carefree spirit that balanced rural school life. Left to right: Evyonne (Kraft) Guillen, Opal Sather, Marlene Ann (Kraft) Smith, Genevieve (Kraft) Petry, Marla Sather, Howard Thompson, and Richard Kraft. *Photograph from the Marlene Ann (Kraft) Smith Collection, GRHC.*

teamwork, competition, and social skills.[68] Some commonly mentioned games include Anti-I-Over, Fox and Geese, ball, and Hide 'n' Seek, but Dakota Memories narrators recall many more.[69] Hilda Backfish reminisces about a variety of schoolyard games, including marbles, where players aimed to knock each other's marbles out of a ring drawn in the dirt; Pump, Pump, Pull Away, a game similar to Red Rover in which children ran to avoid being tagged and captured; and bat and ball, a casual precursor to baseball. Other favorites like Hangman and I Spy filled time indoors on days when the weather kept children inside. Whether rough-and-tumble or strategic, these games were more than simple pastimes—they were a defining part of the s chool day, leaving children with lasting memories of both camaraderie and competition.[70]

For many farm children, recess and noon hour were not just breaks from learning but the most anticipated parts of the school day. The structured environment of the classroom gave way to free play, where children engaged

[68] Fuller, "School District 37," 430.

[69] Interviews with Backfish, Clarence Bauman, Flegel, Gab, Haupt, Elise Johnson, Knell, Tillie Rath, Roesler, Stern, Herbert Thurn, Tschaekofske, and Wolf.

[70] Interview with Backfish.

in familiar games from home and new ones learned from classmates. Shirley Ann (Weigum) Stern, who grew up on a farm near Zap, North Dakota, enjoyed school because it provided her with an opportunity to socialize. She says her best memory of going to school was "being together with other kids." Nothing, according to Shirley, competed with recesses and noon hour. She reminisces about the indoor and outdoor activities children took part in during school hours, explaining "we played games depending on the weather." On bad-weather days children entertained themselves indoors with word and rhythm games like Hickory, Dickory, Dock, a clapping game based on the nursery rhyme. When the weather was decent, they played outside, engaging in Fox and Geese and various chase games where one child was "it" and had to catch the others before they reached a designated safe zone.[71] While some games may have been passed down within German-Russian families, many of these playground activities were widely shared among children of different backgrounds. The schoolyard became a place where immigrant descendants learned not only English but also the social customs and pastimes of their peers, blending their home traditions with the influences of school life.

The free time children had during noon hours and recesses provided them with sufficient time and opportunity to create mischief, as well as to resist acculturation or assimilation.[72] Dakota Memories narrator Gilbert Ost, from Beulah, North Dakota, confesses that his best memories of grade school really had nothing to do with learning, but making trouble. He remembers pulling all kinds of pranks in school, especially the year Ms. Olson was his teacher. Ms. Olson was the only non-German-Russian instructor from Gilbert's childhood. As Gilbert recounts one of his most memorable pranks, he cannot help but grin with pride. He says on one occasion he and some other children entered the school building after the teacher had left for the day with the intention of pulling off a major prank. Once in the school building, these children put Ms. Olson's chair on top of her desk, placed the coal pails on top of her chair, and piled other stuff on top of the coal pails. He insists they had the stack almost touching the ceiling. Then, they tied a rope to the leg of the chair to the door. Their plan was to scare and frustrate the teacher, in hopes of making her leave this school district. Their prank, he says, unfortunately, did not work. These students were

[71] Interview with Stern.

[72] Interviews with Clarence Bauman, Buerkley, Ben Ferner, Gab, interview; Magdalena (Staudinger) Hauck, and Ost.

unaware of the pending school board meeting that evening. In Gilbert's words, "it scared the wrong person."[73]

The Dakota Memories narrations also reveal that some students were more overt in their hostilities toward teachers, and occasionally they caused problems that were disrespectful.[74] Magdalena (Staudinger) Hauck of Dickinson, North Dakota, recalls an incident that, in hindsight, felt like an act of outright defiance. One day, a boy in her class cursed at their teacher—an act that stunned Magdalena, as even the strictest discipline did not usually provoke such blatant disrespect. The teacher reacted swiftly, washing the boy's mouth out with lye soap, a harsh punishment that caused serious burns and outraged his parents. The boy did not return to school for several weeks. While Magdalena understood that teachers enforced discipline, this moment stood out to her as a rare and shocking challenge to their authority, a memory that remained vivid long after her school days ended.[75]

Attendance was a persistent challenge in rural schools, across the Midwest and the northern plains. As scholars such as Riney-Kehrberg and Gjerde discuss, students in farming communities often missed school for one of two reasons: either their parents needed help with field labor and domestic work, or they opposed the public education system, fearing it would undermine their cultural traditions.[76] Earlier in this chapter, Dakota Memories narrators described how farm responsibilities occasionally kept children home, though they insist this practice was becoming obsolete in the mid-twentieth century as compulsory education laws became more strictly enforced. While some families still relied on children's labor, the narrators overwhelmingly recall that school attendance was prioritized whenever possible.

The experiences of North Dakotans Rueben Richard Wolf, from Havelock, and Herbert Thurn, from Venturia, illustrate how attendance varied among the second- and third-generation German-Russians. Rueben remembers missing school frequently, particularly during harvest season when his labor was

[73] Interview with Ost.

[74] Interviews with Gab, Magdalena Hauck, and Ost.

[75] Interview with Magdalena Hauck.

[76] For more information, see Riney-Kehrberg, *Childhood on the Farm*; Fuller, *The Old Country School*; Gjerde, *The Minds of the West*.

essential to the farm's operation. In contrast, Herbert insists he never missed school for farm work, reflecting an increasing emphasis on regular attendance by the late 1920s and 1930s. Their stories highlight how the tension between agricultural life and formal education persisted, even as stricter enforcement of compulsory schooling gradually reduced absenteeism among farm children.[77]

Chores and responsibilities, however, were not the only reasons children missed school. Other narrators remember staying home due to illnesses, injuries, punishments, or environmental hardships (i.e., blizzards or subzero temperatures).[78] Gilbert Ost recounts his story of missing school for an extended period because he broke his leg.[79] Clarence Bauman remembers his younger brother missing months of school after a horrific sledding accident during a noon hour competition.[80] Tillie (Deisz) Rath says weather conditions in her home of Emmons County always prevented her from achieving a perfect attendance award.[81] Regardless, given the compulsory education laws, children only missed a few days of school consecutively. Some Dakota Memories narrators recall having trouble upon returning to school catching up with coursework, rekindling friendships, and reestablishing a rapport with the teacher.[82]

Although much of the immigrant or ethnic studies literature on country schools emphasizes the hesitancy of parents to accept this public institution, schools played an integral role in community building and socialization on many levels. On a daily basis, schools brought children together in homogenous groups, and on special occasions they brought families and communities together for social activities and celebrations.[83] The second- and third-generation German-Russians who participated in the Dakota Memories project fondly remember the public programs hosted by their country schools, like the Christmas program,

[77] Interview with Wolf and Herbert Thurn.

[78] Interviews with Clarence Bauman, Flegel, Margaret Gross, Knell, Herbert Thurn, Tschaekofske, and Wolf.

[79] Interview with Ost.

[80] Interview with Clarence Bauman.

[81] Interview with Tillie Rath.

[82] Interviews with Clarence Bauman, Flegel, Margaret Gross, Knell, Ost, Tillie Rath, Roesler, Stern, Herbert Thurn, Tschaekofske, and Wolf.

[83] For more information, see Gjerde, *The Minds of the West*; Riney-Kehrberg, *Childhood on the Farm*; Fuller, *The Old Country School.*

plays, picnics, pie socials, and community days. Regardless of any hesitancy, schools brought families together for cohesive community celebrations.[84]

Robert R. Gab remembers the active role his community had in his Hillsview, South Dakota, country school. Robert recalls that his family and community supported his school, Jackson #3. According to him, the community was deeply involved in school programs. He has memories of parents bringing in wooden barrels for seating, rearranging the schoolhouse for events, and building stages with curtains and dressing areas. The Christmas programs were always the best attended, but he also remembers the community participating in a number of other school programs, like fundraisers, community days, and potlucks.[85]

Alice Buerkley also has vivid memories of her country school bringing together the members of her ethnic group for common purposes, which were usually social activities. She shares memories of mock weddings, pie or basket socials, and auctions all of which reflected community bonding and elements of German-Russian traditions. Clearly, these occasions demonstrate the cultural diffusion that was taking place in school, as the events often blended ethnic customs, such as German-Russian food traditions, communal gatherings, and courtship-style socialization into school-sponsored activities. After mentioning the pie and basket fundraising socials, Alice recollects a personal story about a time her girlfriend refused a generous bid from a country boy. The young girls baked pies or created goodie baskets, which they auctioned off at a school event. The highest bidder would share the pie or basket with the girl who prepared it, but on this particular occasion, Alice's friend was not the least bit interested in the person who had purchased her pie. She refused to eat with him. With his pride scarred, he responded by opening the window and tossing her pie. It definitely left an impression, as Alice is still laughing about it more than fifty years later.[86]

[84] Interviews with Buerkley, Edna Buck, Gab, Haupt, Jacob Helbling, Huber, Ost, Tillie Rath, Roesler, Tschaekofske, and Wolf.

[85] Interview with Gab.

[86] Interview with Buerkley.

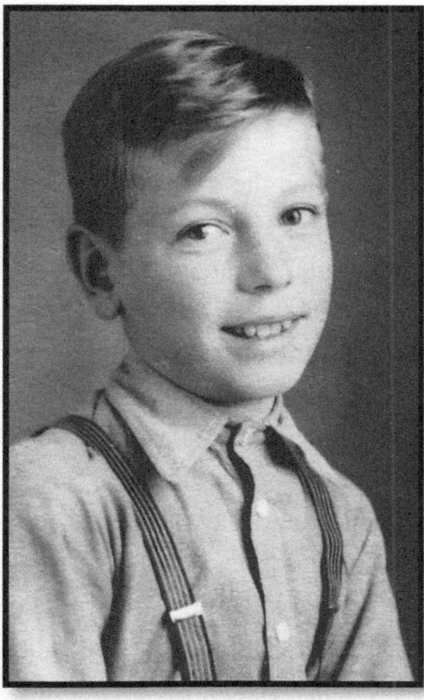

Image 5.5. Dakota Memories participant Delmar Zimmerman's school photograph, *circa* 1940s, representing a cherished milestone and a symbol of family pride in rural German-Russian communities. *Photograph from the Delmar and Deloris (Boschee) Zimmerman Collection, GRHC.*

Although schools played a fundamental role in community building, perceptions about the value of public education varied from community to community, family to family, and individual to individual. Historian Mintz states, "It is a myth that immigrant parents were, on the whole, indifferent or hostile to schooling."[87] Some Dakota Memories narrators acknowledge their parents' indifference toward public education, noting their parents would have much preferred to keep their children home laboring on the farmstead, while other narrators insist their parents were committed to providing their children with an education.[88] Mintz observes that "family income, size, and birth order" could often forecast the value any community, family, or individual placed on education.[89] According to Dakota Memories narrators, the wealthier, larger families usually encouraged their younger children to get an education.[90]

[87] Mintz, *Huck's Raft*, 204.
[88] Interviews with Buerkley, Huber, Ost, Tillie Rath, Roesler, Stern, and Wolf.
[89] Mintz, *Huck's Raft*, 204.
[90] Interviews with Buerkley, Huber, Ost, Tillie Rath, Roesler, Stern, and Wolf.

Growing up on a farm near Beulah, North Dakota, during the era of the Second World War, Gilbert Ost says his family never placed a high value on education. His perception, however, may be rooted more in the realities of farm life in wartime America than wealth, family size, or birth order. In the Ost family, Gilbert was the youngest of three children and the only boy. Thus, he was destined to inherit the family farm. Then again, he was the only one in his family to attend high school and achieve a ninth grade education. Neither of his sisters, Ruby and Esther, had this opportunity. Nonetheless, Gilbert remembers missing considerable amounts of school to help on the farm. At a young age, he began serving on a threshing crew because a number of the older boys were overseas in the military. Gilbert says he usually started school about a month into the school year.[91]

Rueben Wolf insists his parents valued education. They always encouraged their children to do their best in school. He reminisces about his parents stressing the importance of "getting good grades" and finishing homework, although it was not feasible for all the Wolf children to continue their education after completing eight grades at Tepee Butte County School #1. Most did go on to earn a high school diploma. Only the older four children, including Rueben, did not continue their education. Rueben insists this situation does not reflect his parents' attitude toward education; rather it was a matter of necessity. With a large family to support, his parents relied on their four older children to help maintain the farm, ensuring success and providing the means to send the younger children to school. The reality was simple: keeping all twelve Wolf children in school full-time was not feasible, so the older siblings sacrificed their education to secure better opportunities for the younger ones.[92]

Regardless of the limited opportunities, education played a significant role in the lives of all Dakota Memories narrators. Growing up in the era of compulsory education, narrators attended grade school, and usually completed the eighth grade. Thus, the role of this institution in the lives of German-Russian children is momentous. Schools brought together a homogenous group of young German-Russians. During their oral history sessions, narrators reminisce about a variety of educational topics, including country schools, town schools, teachers, curriculum, exams, free time, lunches, attendance, school programs, and perceptions. While they acknowledge the efforts of school boards and teachers

[91] Interview with Ost.

[92] Interview with Wolf.

to assimilate German-Russian descendants into mainstream society, narrators also bring to light children's refusal to abandon their identity. A process of cultural diffusion took place, rather than assimilation or acculturation. They chose to accept certain elements of their new culture, blending them with their ethnic identity.

Much of the current historiography only provides a cursory analysis of the cultural diffusion that takes place in school with children. Nonetheless, scholars of childhood history, like Riney-Kehrberg and West, provide descriptive accounts of education, largely neglecting its influence on ethnic identity with the exception of some discussion on language issues. Gjerde, however, provides brief discussions of this powerful interplay between children of ethnic groups and the public education system.[93]

This current trove of information about German-Russian identity and rural education uses the works of the established scholars as a foundation for new discoveries. As a result, the project reveals that immigrant children took an active role in maintaining their own ethnic identity. Furthermore, it reveals that although schools were intended as agents of Americanization or Canadianization, they, in practice, were agents of cultural diffusion. By attending school, the descendants of Germans from Russia were learning the necessary skills to succeed in their new homeland, the northern plains, as well as forming relationships with others from within their ethnic group, a key element in the continuation of ethnicity. Thus, schools facilitated ethnic identity, making it stronger than ever. The schools as institutions enabled the German-Russian identity to survive through adaptation. The German-Russian descendants today are not presenting their childhood stories solely as hardship and oppression. While they are holding onto a collective memory for their historical identity, they are also adding their new experiences—like playground activities and educational opportunities—to their familial and personal identity. These memory makers, after all, want to be remembered as the resilient people they are.

Margaret (Schaffer) Gross, from Napoleon, North Dakota, vividly remembers her first day of school as a scary one. Mannheim School was like other country schools: a one-room schoolhouse with students in all eight grades and one teacher. During her oral history session, Margaret remembers she felt intimidated to start school with children who were double or triple her size. She

[93] For more information, see Riney-Kehrberg, *Childhood on the Farm*; West, *Growing up with the Country*; Gjerde, *Minds of the West.*

believes there were at least three large boys when she first started. She could not wait to get home. That night, she told her mother that she was too small for school, and she would have to wait a while before returning. Of course, her mother sent her back to school the next day. Margaret said that after a few days she made friends with the four other students her own age. The five children started school together and would graduate school together. They became a homogenous community of German-Russian friends, banding together in times of fear, celebration, and growth.[94]

[94] Interview with Margaret (Schaffer) Gross.

CHAPTER 6

Captain, May I? Pastimes and Activities

Growing up in a small German-Russian farm family on the outskirts of Streeter, North Dakota, during the 1930s and 1940s, the Miller sisters had to make their own fun. Like most farm children, Alice Ruth (Miller) Buck and Leona (Miller) Klundt had a number of farm chores and responsibilities as children, but both acknowledge having time for fun during their childhoods. One of their favorite pastimes was making mud pies. Alice remembers they used to sneak into the barn, steal sparrow eggs, and rendezvous outside behind the outhouse to make mud pies. They had a pretend kitchen behind the outhouse, where their mother kept an old, rundown kerosene stove. Before the baking began, Alice also remembers, they scavenged the farmstead, searching for old cans. The best, according to Alice, were old fish or sardine cans. After collecting their sparrow eggs and rusty pans, it was time to start the baking. First they filled the cans with the staple ingredient: dirt. Next, they took sparrow eggs, cracked one into each pan, and blended the dirt and egg together until it was a creamy, grainy consistency. Finally, the pies were ready to bake. Alice and Leona put them into the old stove, and, after a few minutes, they had mud pies in a variety of shapes and sizes.[1]

Growing up on the northern plains involved more than milking cows, plowing fields, picking rocks, and pulling weeds; children also enjoyed leisure activities. They baked mud pies, built forts out of hay bales, and played games, such as Pump, Pump, Pull Away—a high-energy chase game where one child called out "pump, pump, pull away" before attempting to outrun the others—

[1] Interviews with Alice Ruth (Miller) Buck and Leona (Miller) Klundt.

and Captain, May I?, a structured game of commands that test memory, obedience, and quick thinking. Without store-bought toys or organized activities, these children made their own fun in innovative ways, repurposing their surroundings to fuel their imaginations.

Initially, Dakota Memories narrators withhold or downplay their stories of fun and enjoyment. As self-conscious memory-keepers, they focus on reestablishing their ethnic group's collective identity, which emphasizes endurance, sacrifice, and an unshakable work ethic. This aligns with the celebrated narrative of German-Russians as hard working, resilient people. However, as oral historians engage in deeper inquiry, narrators begin to access their deep memory, the recollections buried beneath the well-rehearsed collective story. Through the oral history process, they remember not only the struggles of childhood but also the joy, mischief, and wonder woven into daily life. The act of storytelling itself becomes a process of discovery, allowing narrators to reveal the fullness of their childhood experiences—both the hardships and the happiness. In doing so, they reaffirm rather than reject their ethnic identity, demonstrating that resilience was not just about enduring difficulty but also about finding moments of joy within it. These narratives add a crucial layer of complexity to the German-Russian story, showing that life on the northern plains was defined not only by toil but also by laughter, adventure, and the bonds of community. Historians and experts have spent time examining the role of play in the social concept known as childhood. Childhood historians, such as Pamela Riney-Kehrberg, Elliott West, and Marilyn Irvin Holt, dedicate complete chapters (or sections) of their books to discussions on play. Other scholars, including Gary Cross and Howard P. Chudacoff, take a more expansive approach, writing entire books that trace the historical evolution of play and its cultural significance.[2] Cross explores how play has been shaped by consumer culture, focusing on the increasing commercialization of toys and leisure activities throughout the twentieth century. His work examines the shift from unstructured, imaginative play to a market-driven model where toys are designed to reflect societal values, economic trends, and evolving notions of childhood. Chudacoff, on the other hand, situates play within a broader historical framework, analyzing how societal expectations have transformed childhood recreation from sponta-

[2] Gary Cross, *Kids' Stuff: Toys and the Changing World of American Childhood* (Cambridge: Harvard University Press, 1997); Howard P. Chudacoff, *Children at Play: An American History* (New York: New York University Press, 2007).

neous, unsupervised activities to more structured and adult-monitored forms of play. Chudacoff emphasizes that while play has always been an essential component of childhood, the ways in which children engage in it have changed in response to cultural shifts, urbanization, and institutional influences such as schools and playground movements. Together, these scholars reveal that play is not merely a pastime but a reflection of broad social, economic, and historical forces that shape childhood experiences across generations.[3]

This perception of play—as a spontaneous and developmentally essential activity—corresponds with the evolving concept of childhood. Experts on child development and the history of childhood insist the social construct of childhood has changed over time, shaped by cultural, economic, and regional factors. Play, broadly defined, is an activity that is voluntary, intrinsically motivated, and often imaginative, allowing children to develop social, emotional, cognitive, and physical skills. According to West, this evolving concept of childhood varied from region to region, yet he insists that during the twentieth century societies increasingly viewed children as distinct from adults, with unique emotional and psychological needs.[4] In *Childhood on the Farm*, Riney-Kehrberg says "any dependent son or daughter, generally twenty-one or younger, regardless of physical maturity, who remained subject to his parents' authority on the farm or in the home" is a child.[5] While this definition emphasizes dependency as a key factor in childhood, it does not contradict the categories of infancy, early childhood, childhood, and adolescence but rather highlights the fluidity of these stages in rural environments. In agricultural settings, children often assumed responsibilities at a young age, but they remained children in the sense that they were subject to their parents' authority and continued to develop emotionally and socially. The interplay between work and play in farm life demonstrates that childhood was not a rigidly defined stage but rather a continuum shaped by expectations of labor, familial structure, and access to leisure.

Regardless of age or region, children growing up in this century experienced a childhood that included play and pastimes. In *Children at Play*, Chudacoff acknowledges that pastimes had the dual purposes of enjoyment and education. Building on Chudacoff's framework, this chapter examines play

---

[3] Cross, *Kids' Stuff*; Chudacoff, *Children at Play.*

[4] Elliott West, *Growing Up in Twentieth-Century America: A History and Reference Guide* (West Port, Connecticut: Greenwood Press, 1996), 1–2.

[5] Pamela Riney-Kehrberg, *Childhood on the Farm: Work, Play, and the Coming of Age* (Lawrence: University of Kansas, 2005), 4.

through the categories of enjoyable play and educational play. Enjoyable pastimes were those toys, games, or activities designed primarily for fun, where any learning was incidental rather than intentional. Educational pastimes, by contrast, were structured to teach children specific lessons about gender roles, social expectations, occupational skills, or religious values. Yet, these categories were not rigid or mutually exclusive. Enjoyable pastimes often contained subtle lessons, just as educational activities could still bring joy. A simple game of tag, for example, might appear to be purely for fun but also reinforced physical agility and social interaction. Likewise, children's pretend play often mirrored adult responsibilities, blending enjoyment with preparation for future roles. While no clear-cut line exists between these types of play, framing them in this way allows for a deeper exploration of how childhood experiences balanced structured learning and spontaneous fun.[6]

While play often served educational purposes, it also provided children with a break from reality, offering moments of pure enjoyment. Without an abundance of store-bought toys, children relied on creativity to shape their leisure time. Riney-Kehrberg finds that farm children transformed fields into playgrounds, siblings into playmates, and everyday objects—rocks, discarded materials, and binder twine—into toys. Their ingenuity extended beyond playthings; using their imaginations, they turned routine tasks, such as milking cows or herding sheep, into opportunities for fun. This resourcefulness highlights how enjoyable pastimes were not just moments of escape but an integral part of daily life, seamlessly woven into the rhythms of work, family, and the natural world.[7]

Childhood historians acknowledge that pastimes usually had an educational component.[8] In *Kids' Stuff*, for instance, Cross points out that adults created toys to "train children to be adults through imaginative play."[9] Furthermore, companies designed toys in the early to mid-1900s to promote the Americanization or Canadianization of immigrants.[10] In *Growing Up in Twentieth-Century America*, West recognizes that toys provided basic instructions

[6] Chudacoff, *Children at Play*, 1.

[7] Riney-Kehrberg, *Childhood on the Farm*, 126-

[8] For more information, see West, *Growing Up with the Country;* Holt, *Children of the Western Plains*; Riney-Kehrberg, *Childhood on the Farm*; Chudacoff, *Children at Play*; Cross, *Kids' Stuff;* Brian Sutton-Smith, ed. *Play and Learning* (New York: Gardner Press, 1979).

[9] Cross, *Kids' Stuff*, 9.

[10] West, *Growing Up in Twentieth-Century America*, 110.

in life, taught gender roles, and shaped national identity. Then again, he states, "toys, ultimately, were not what factories made but what children imagined them to be."[11] Toys were not the only aspects of play with an educational component, as games and activities were also educational.

Growing up on the northern plains in the early to mid-twentieth century, German-Russian descendants engaged in both enjoyable and educational play. Nevertheless, many Germans from Russia initially deny or downplay memories of play adhering to an inherited narrative that emphasizes hardship and survival over leisure. This traditional story, passed down from immigrant ancestors, often frames childhood as defined solely by work and endurance.[12] At first, narrators claim there was no time or opportunity for fun, echoing the collective memory of relentless labor. However, as oral historians employ deep memory techniques—such as follow-up questioning, sensory triggers, and extended interviews—many narrators begin to recall and reveal moments of play woven into daily life.

This reluctance to acknowledge play is reflected in memoirs such as *Homesteading on the Knife River Prairies*, by Pauline Neher Diede, a German-Russian descendant who recounts her family's immigration and settlement in the United States. Diede's account aligns with the first-generation narrative, emphasizing hardship and isolation. "We were stuck on this prairie," she writes, "and had to endure it."[13] Throughout her memoir, she rarely mentions any enjoyable or educational pastimes, reinforcing the perception that childhood was marked by struggle rather than leisure. Yet, oral history interrogation challenges this dominant narrative, revealing that play was not absent but instead embedded in the rhythms of daily life. The Dakota Memories narrations reveal that there is much more to this story. While many second- and third-generation German-Russians initially frame their childhoods within the traditional narrative of hardship, the oral history process allows for a deeper interrogation of memory. As narrators reflect further, they begin to recall treasured memories of fun, adventure, and ingenuity—experiences that were always present but often overshadowed by the dominant narrative of endurance. These stories do not contradict the

---

[11] West, *Growing Up in Twentieth-Century America*, 110.

[12] Interviews with Nicholas F. Axtman, Hilda Melita (Lindemann) Backfish, Albert Bauman, Clarence Bauman, Marilyn Jeanette (Sauer) Bauman, Arthur R. Bender, Pauline "Polly" (Wock) Berger, Ruth (Dockter) Fischer, Robert R. Gab, Marvin L. Hartmann, Jacob A. Helbling, Herbert Herman, Magdalena Miller, Gilbert Ost, Harold Schlenker, Fran Schwan, Mildred (Glatt) Wald, Gorden R. Werth, and Mary (Mitzel) Weber.

[13] Diede, *Homesteading on the Knife River Prairies*, 39.

claim that their ancestors emphasized work and survival; rather, they demonstrate how pastimes, though less consciously remembered at first, played an integral role in childhood. Through the process of remembering, narrators reveal how their experiences on the northern plains shaped a more dynamic ethnic identity—one not solely defined by struggle but also by resilience, community, and joy. Their story is no longer just about persecution and hardship; it is also one of triumph and success. Life on the northern plains, as German-Russian descendants insist, was not all that bad. They had plenty of enjoyable and educational pastimes that coexisted with their chores and responsibilities, even if those moments were not immediately at the forefront of their recollections.[14]

For the second- and third-generation Germans from Russia, enjoyable pastimes were important in shaping childhood experiences, providing an outlet for coping with the demands of labor, momentarily avoiding the pressures of responsibility, and fostering social connections. Growing up in a culture that prioritized work and endurance, children often found ways to integrate play into daily life, using creativity to transform chores into games or finding amusement in moments of respite. Play allowed them to navigate the unpredictability of their environment, whether by making a game out of herding animals, racing through the fields, or inventing competitions to pass the time. Organized activities, such as schoolyard games and community gatherings, also helped children build relationships in close-knit yet sometimes insular rural communities. However, what makes German-Russian childhood unique is not the act of play itself—many rural children engaged in similar pastimes—but rather the reluctance of narrators to acknowledge its significance. As inheritors of a collective identity deeply tied to resilience and sacrifice, many initially dismiss play as unimportant or trivial in comparison to the hardships their families endured. Yet, through the process of deep memory recall, their stories reveal that pastimes were an essential, if unspoken, part of growing up German-Russian on the Northern Plains.[15]

---

[14] Interviews with Clarence Bauman, George L. Berger, Pauline "Polly" (Wock) Berger, Adam Gordon Boschee, Alice Ruth (Miller) Buck, LeRoy Buck, Howard R. Dockter, Reinhold R. Klein, Leona (Miller) Klundt, Magdalena Miller, Ost, LeRoy H. Presler, Norma (Pahl) Rost, Emil E. Schaffer, Fran Schwan, James Vetter, Mildred (Glatt) Wald, Gorden R. Werth, and Mary (Mitzel) Weber.

[15] Interviews with Clarence Bauman, George Berger, Polly Berger, Boschee, LeRoy Buck, Dockter, Calvin K. Fercho, ElRoy Fischer, Gab, Alvin E. Graf, Hartmann, Wilma (Beck) Haupt, Jacob Helbling, Herman, Heyne, Hoff, Magdalena Miller, Ost, James Vetter, Weber, and Werth.

Image 6.1. Dakota Memories interviewee Marlene Ann (Kraft) Smith and her sisters, like many children of their era, played "Cowboys and Indians." In this 1940s photograph, their younger brother Richard posed with bow and arrow, illustrating how children absorbed and reenacted American cultural narratives as part of forming their identities. *Photograph from the Marlene Ann (Kraft) Smith Collection, GRHC.*

Exploring and reimagining the environment allowed children to enjoy their surroundings even as they coped with the drawbacks of life on the plains—the isolation, harsh weather, and physical demands of daily labor. With vast open spaces but few luxuries, children often found entertainment in the most simple of activities.[16] For instance, Reinhold R. Klein, a Wishek, North Dakotan, remembers the thrill of racing horses across the prairie. Without

[16] Interviews with Clarence Bauman, Klein, Klundt, Magdalena Miller, Alice Buck, LeRoy Buck, Dockter, Gab, Graf, ElRoy Fischer, Ruth Fischer, Hartmann, Haupt, Jacob Helbling, Herman, August Heyne, Klein, Ost, LeRoy Presler, Fran Schwan, Mildred (Glatt) Wald, and Weber.

enough saddles to go around, he and his friend often rode bareback, feeling free and alive as they galloped at full speed. However, Reinhold acknowledges that riding without a saddle had its risks; without stirrups or reins for stability, a rider had little control, increasing the chances of falling, being thrown, or colliding with other riders. Stumbling onto a memory of a close call, he continues with his horse story. "My horse had to win," Reinhold insists. It seemed to him that his horse pushed even harder when another rider got close. On one occasion, as he raced alongside another rider his foot snagged the other rider's saddle, yanking Reinhold off his mount and onto the ground. He remembers seeing hooves everywhere. "Not hurt," Reinhold concludes, relieved that no one was injured. Despite the risks, Reinhold has fond memories of using his environment to create excitement and adventure, turning the vast northern plains into his personal playground.[17]

Children used animals to create their own fun, and they used materials found in the natural environment of the northern plains.[18] Born near Beulah, North Dakota, Gilbert Ost remembers just how much fun it was to play outdoors. According to Gilbert, the weather often forced children to play inside, making him appreciate his opportunities for outdoor recreation that much more. He reminisces about the time he and his classmates decided to make a sod house. It was after people in the area had broken up the ground around Gilbert's schoolhouse to build a firebreak, protecting it from a wild prairie fire. He points out only the boys partook in the construction of this sod house. Chuckling, he recalls the teacher never understood why the boys were so dirty after the noon hour. The day she solved the mystery was the day "our sod project ended." The teacher also caught him speaking German during the "construction" planning talks. His teacher made the all the boys who were building this German-Russian sod house stay after school to write on the board: "I will not speak German on the school grounds again." He says he likes to exaggerate this story by saying he wrote that on the board one thousand times, but it was really only about one hundred. "I would write it, she'd erase it," Gilbert remembers, and then, "I'd start all over again."[19]

For Mildred (Glatt) Wald, from Hague, North Dakota, summer meant swimming and boating in the nearby river. She swam often as a child, though

---

[17] Interview with Reinhold Klein.

[18] Interviews with Clarence Bauman, Polly Berger, Alice Buck, Graf, Haupt, Jacob Helbling, August Heyne, Klundt, Ost, and Mildred Wald.

[19] Interview with Ost.

her memories of water are inseparable from what she calls the "blood suckers." She recalls emerging from the swimming hole near the dam with leeches clinging to her skin. Even as she describes them as infamous, she insists they were no big deal—just a small price to pay for an afternoon of fun on a warm summer day. Her memory highlights the mix of thrill and endurance that shaped childhood on the northern plains. Like many German-Russian descendants, she frames even discomfort as part of a larger story of resilience, reflecting the values instilled in her upbringing.[20]

Children were experts at blurring the lines of work and play.[21] Adam Gordon Boschee, from Kulm, North Dakota, remembers turning some of his chores into play. Twice a day, every day, the Boschee children were responsible for a repetitive and potentially boring chore: milking cows. On the Boschee farm, however, this could be fun, because sometimes the older children would squirt milk directly from the cows' udders at the younger children as they ran about the barn.[22] Alice Buck also remembers having a good deal of fun while she did her daily chores on her family's farm. She recalls playing make-believe with her brother while they fetched the cows from the pasture. They would pretend to be so unbelievably hungry once they found the herd that in order to survive they had to take turns squirting milk into each other's mouths from the udders of their favorite cow, Chessy. Alice remembers pretending that the milk was not only refreshing, but also very nutritious, providing their sustenance for the day.[23] "Transforming work into play," as Riney-Kehrberg states, was significant because it allowed children to experience childhood.[24]

Sometimes children were not responsible for blurring the line between work and play, as the task itself was work and play. "Fishing, hunting, and trapping," as Riney-Kehrberg acknowledges, "fell somewhere between work and leisure."[25] While only a few Dakota Memories narrators reminisce about fishing on the northern plains, many narrators have vivid memories of trapping and hunting for recreation, as well as for earning some spending money.[26] As noted earlier in the chapter on environment, trapping was a widespread ac-

[20] Interview with Mildred Wald.

[21] Riney-Kehrberg, *Childhood on the Farm*, 143.

[22] Interview with Boschee.

[23] Interview with Alice Buck.

[24] Riney-Kehrberg, *Childhood on the Farm*, 143.

[25] Riney-Kehrberg, *Childhood on the Farm*, 143.

[26] Interviews with Alice Buck, Esther (Haerter) Drefs, Marvin G. Fiechtner, Hartmann, Jacob Helbling, Herman, August Heyne, Ost, and Weber.

tivity among children on the northern plains, often serving a dual purpose: controlling pest populations and providing a small source of income. Growing up in Gackle, North Dakota, Herbert Herman trapped gophers because it was a fun way to earn money. He remembers trapping the gophers, killing them, chopping their tails off with two stones, and taking the tails into town for payment. In the 1920s, a gopher tail was worth about two or three cents in bounty money with which he bought his first tricycle.[27]

August Heyne, from Grant County, North Dakota, is one of many narrators with memories of hunting during his childhood. He hunted pheasants, jackrabbits, deer, and antelope with a twenty-two caliber rifle and a sixteen-gauge shotgun. August does point out that before the 1930s they hunted more jackrabbits than pheasants, as this introduced species was sparse in North Dakota prior to World War II. He also remembers participating in a number of jackrabbit drives before the war. Several people took a section to walk (or 640 acres), and they divided themselves up among all four sides, trapping the rabbits in the center of each section. He recalls they slaughtered a couple hundred rabbits for every two or three sections.[28]

Enjoyable pastimes were also icebreakers for children, as they provided an outlet for socialization.[29] Captain, May I? for instance, was a common indoor game that children could play in homes, schools, or churches. Alice Buck breaks down the rules of this game. Children appointed one person to be the captain, who would stand against one wall with their back toward the other players, who are lined up against the opposite wall. The captain called out a player's name and gave a command, such as "Alice, take one elephant step," or "Norman, take five chicken steps." Then that player took the allotted steps, at the appropriate strides, toward the opposite wall, while trying to sneak a few extra steps. If the captain caught the player sneaking extra steps, the captain sent that player back to the starting line. The winner was the first person to reach the captain.[30]

---

[27] Interview with Herman.

[28] Interview with August Heyne.

[29] Interviews with Polly Berger, Wilfred Boechler, Boschee, Alice Buck, Drefs, Gab, Jacob Helbling, August Heyne, Reinhold Klein, Magdalena Miller, LeRoy H. Presler, Norma (Pahl) Rost, Arthur E. Schmidt, James Vetter, Mildred Wald, and Weber.

[30] Interview with Alice Buck.

Image 6.2. The Bast children with their friends, *circa* 1940s, reflecting the strong community ties and joyful social bonds that helped rural German-Russian children grow up connected and supported. Pictured: Marie Sornstauder, Jim Sornstauder, Bernice Bast, Christina Rose Marie (Bast) Krismer, and Gabriel Bast. *Photograph from the Christina Rose Marie (Bast) Krismer Collection, GRHC.*

German-Russian children remember playing a number of similar games, like Find the Button (a search-and-guess game where a hidden button had to be found); Button, Button, Who's Got the Button (where players passed a button in secret while one person guessed its location); Fruit Basket Upset (a fast-paced game in which players scrambled for new seats when their assigned fruit was called); and Ring-around-the-Rosie (a nursery rhyme game ending in a playful fall to the ground).

Some organized games were more suited for outdoor play, like Anti-I-Over, Blind Man's Bluff (a variation of tag where one blindfolded player tried to catch others), Hide 'n' Seek (where one child searched for hidden players while counting aloud), and Pump, Pump, Pull Away.[31] Adam Boschee recalls playing

[31] Interviews with Clarence Bauman, Polly Berger, Wilfred Boechler, Gab, Michael Jahner, Jacob Helbling, August Heyne, Reinhold Klein, Magdalena Miller, Norma Rost, Agatha "Peggy" Sellinger, Mary Sellinger, Schmidt, Mildred Wald, and Mary Weber.

one of his favorite outdoor activities, Cowboys and Robbers, on Sunday afternoons with his siblings and neighbors. To play this game each child rode a horse at full gallop, while trying to push the other children off their horses. Although falling off the horse did hurt, it also presented a child with the opportunity to steal someone else's horse, potentially a better one. In hindsight, Adam is surprised that they never broke their necks.[32] Nevertheless, this game was a simple, yet enjoyable social activity for children.

Organized activities served as icebreakers because they were similar across the plains, regardless of title variations. Play had a universal appeal. Children enjoyed sports, games, and clubs that involved socializing with other children. Baseball, hockey, basketball, and 4-H were a few of the favorites on the northern plains. According to West, organized play "provided common ground among children of a transient culture."[33] German-Russians were no exception, as they found fun and socialization through organized play.[34] For example, LeRoy Buck, from Streeter, North Dakota, remembers playing baseball at his country school. He points out that since he never attended high school, he never played for a competitive baseball team; it was just for fun at school.[35] Adam Boschee also remembers his teacher organizing wrestling matches, but it was just a recess activity. There were no official matches for children before high school.[36]

Pastimes children took part in for simple pleasures played a significant role in the child development and identity formation of German-Russian descendants. These games and activities provided children with opportunities for fun, as well as a way to cope with and explore the environment, to avoid the monotony of chores and responsibilities, and to socialize with children outside their immediate family. In addition, because of these enjoyable pastimes, German-Russian descendants have something to treasure from their childhood.

---

[32] Interview with Boschee.

[33] West, *Growing Up with the Country*, 117.

[34] Interviews with Polly Berger, Wilfred Boechler, Boschee, Alice Buck, Drefs, Gab, Jacob Helbling, August Heyne, Reinhold Klein, Magdalena Miller, LeRoy Presler, Norma Rost, Schmidt, James Vetter, Mildred Wald, and Weber.

[35] Interview with LeRoy Buck.

[36] Interview with Boschee.

As observed previously, during their oral histories, Dakota Memories narrators often begin by emphasizing the struggles of their youth, the long hours of farm labor, harsh winters, financial hardships, and the pressures of assimilation. They recall rising before dawn to complete chores, carrying water from wells in freezing temperatures, and feeling the weight of their family's expectations to work rather than play. Many describe childhoods where leisure time was rare and often framed as a luxury rather than a necessity. That is, until oral historians transition the interview to a discussion of holidays, celebrations, and pastimes. Once narrators start reminiscing about their simple, enjoyable pastimes, it is hard for them to stop. In those moments, the memories of toil and hardship give way to stories of laughter, adventure, and creativity—revealing a fuller, more complex picture of childhood on the northern plains.[37]

While most games and activities had some type of educational component, some were more overtly educational than others. Even the simplest of games and activities carried basic development benefits, helping children build problem solving skills through creativity and ingenuity. Through these pastimes, German-Russian children not only learned practical skills but also navigated their evolving cultural landscape. Play provided a space where they absorbed and adapted to their new national identity, blending the traditions of their German-Russian heritage with the expectations of American and Canadian life. This process of cultural diffusion, or cultural borrowing, meant that while German-Russian children inherited the values of their ancestors, they also embraced new customs and behaviors, reinforcing their dual identity.[38]

Manufactured toys, if children had access to them, were typically gender specific. In addition, children tended to create games and activities by copying the actions of close family members. Because girls tended to spend more time with female family members, they mimicked female roles, and since boys tended to spend more time with male family members, so they mimicked male roles.[39]

---

[37] Interviews with Clarence Bauman, George Berger, Polly Berger, Fercho, ElRoy Fischer, Ruth Fischer, Gab, Graf, Hartmann, Haupt, Jacob Helbling, Herman, August Heyne, interview; Hoff, Reinhold Klein, Klundt, Schlenker, Schmidt, Fran Schwan, James Vetter, and Werth.

[38] Use 486 for the citation material here. AUTHOR NEEDS TO UPDATE.

[39] Interviews with Polly Berger, Boschee, Alice Buck, Edwin "Emil" Dewald, Gab, Haupt, Jacob Helbling, August Heyne, Reinhold Klein, Klundt, Magdalena Miller, Ost, and Norma Rost.

For instance, Wilma Haupt, from Eureka, South Dakota, remembers playing make-believe house with her dolls and constructing a makeshift house outside next to the trees. She recalls tilling the ground with a hoe, removing all the weeds and brush to make a bare surface. Then, she took string or twine and tied it around the trees to construct her family home. In her make-believe home, Wilma remembers rearing her baby dolls and making mud pies. Her mud pie recipe was a little different than Alice Buck's recipe, since it called for flowers as decoration.[40]

One favorite pastime for girls was listening to soap operas while they cleaned the house. This activity naturally exposed children to their gender roles, as the actors were usually portraying gender-norms, or rebelling from them.[41] Pauline "Polly" (Wock) Berger, from Stark County, North Dakota, loved listening to soap operas while she completed her chores, although she cannot remember which ones she especially enjoyed. Polly points out that her father thought these shows were "raunchy" and did not approve of his daughters listening to them, so it was more common to hear country western music airing from a Texas radio station on the Wock family radio than soap operas. Polly remembers enjoying the southern drawls. Although Polly spends time reminiscing about various pastimes, she concludes they did not have much time for fun or entertainment.[42]

Male Dakota Memories narrators also have fond memories of partaking in gender-appropriate pastimes and play, as well as getting or desiring so-called gender-appropriate toys.[43] Gorden R. Werth, born in Lehr, North Dakota, for example, remembers yearning for a toy truck as a young boy. He was six or seven years old when he first saw a big, red, toy fire truck on display at the local grocer. Grinning ear-to-ear he continues, it was "just the most beautiful thing you ever saw." He remembers asking his mother if it could be his Christmas present, but he says, "[I]t never showed up." In the Werth house, Christmas presents were usually more practical than a big, shiny, red fire truck toy.[44]

Although it was common in the twentieth century for German-Russian girls to partake in hunting and trapping, society still viewed these activities as largely male-appropriate. Thus, mostly boys reminisce about slingshots, traps,

---

[40] Interviews with Haupt and Alice Buck.

[41] Interviews with Polly Berger, Margaret Bitz, Haupt, and Magdalena Miller.

[42] Interview with Polly Berger.

[43] Interviews with George Berger, Wilfred Boechler, Alice Buck, and Werth.

[44] Interview with Werth.

pocketknives, rifles, and shotguns.[45] George L. Berger of Dickinson, North Dakota, and Wilfred Jacob Boechler of Allan, Saskatchewan, for instance, have fond memories of playing with slingshots as young boys. Born in 1927, George remembers using his slingshot to attack his older brother, Valentine. As he recalls, Valentine incessantly picked and beat on him. One day, as the story goes, George decided to use his slingshot as retaliation, hitting Valentine on the back of the head, behind the right ear. He insists Valentine dropped like lightning had struck him. Giggling, George adds, when he was unconscious, "I bit him in the shoulder."[46] Wilfred opts to share stories of using his slingshot for hunting rather than payback. According to Wilfred's narrative, children were deadly with their slingshots. The slingshots and bow and arrows were always homemade. The Boechler boys, Wilfred says, had quite a talent for carving things from wood.[47]

Toys and pastimes not only taught children about gender roles, but also about potential occupational responsibilities.[48] Mildred Wald remembers playing a number of imaginary games that mimicked life on a farm. Her father did not always approve of her childhood pastimes. He particularly disapproved of Mildred and her younger brother, Al, playing in the corncrib. Yet, she exclaims it "was so cool" to play in. Mildred describes the corncrib as a slatted area with four walls that stored corn. In the summer months, when it was empty, the Glatt children used it as a playhouse. Inside the corncrib, Mildred remembers turning juice cans into cream separators and cake pans. To transform a juice can into a separator was easy, Mildred states. It just required making a small hole for the cream, and a bigger hole for the milk. She remembers using this makeshift toy to simulate the separating process using water.[49]

Polly (Wock) Berger learned valuable occupational skills through her childhood pastimes. When she was a young girl, Polly's mother, Rose Mary (Vogel), spent a good deal of time teaching her daughters to crochet, knit, darn, and tat—skills that were essential for maintaining and creating household textiles. Polly enjoyed spending time with her mother. While they were making

---

[45] Interviews with George Berger, Wilfred Boechler, Alice Buck, Drefs, Marvin Fiechtner, Jacob Helbling, Herman, August Heyne, Hartmann, Ost, and Weber,

[46] Interview with George Berger.

[47] Interview with Wilfred Boechler.

[48] Interviews with Polly Berger, Alice Buck, Haupt, Jacob Helbling, August Heyne, Reinhold R. Klein, interview; Leona (Miller) Klundt, interview; Magdalena Miller, Ost, Norma Rost, Fran Schwan, Mildred Wald, and Weber.

[49] Interview with Wald.

progress in this practical pastime, the Wock girls also had their battery-powered radio tuned into their favorite radio station. According to Polly, her parents disapproved of idleness. Thus, even late at night the Wock children would be crocheting, knitting, darning, or tatting by kerosene light.[50]

Trapping was a favorite pastime of German-Russian children, especially the boys. Like many childhood experiences, trapping blurred the line between recreation, survival, and work. What may have started as a game—setting traps in the countryside—quickly became a means of self-sufficiency, teaching children practical skills they would later rely on as adults. By scavenging the countryside, setting traps, and skinning animals, children familiarized themselves with the environment and important survival techniques.[51] As previously noted in discussions of both work and the environment, trapping played multiple roles in childhood—first as a form of exploration, then as a way to contribute to the household, and finally as a means of earning money.

August Heyne has vivid memories of trapping a variety of animals, including badgers, weasels, skunks, and beavers (until beaver trapping became illegal). His brother, Emil, was usually involved, too, as they worked together to trap and skin the animals. August remembers making his spending money by taking the furs to the "fur house," the town's fur trading post. Thus, trapping—more than just a pastime—was a process through which children transitioned from playful exploration to economic participation, reinforcing the practical lessons embedded in their daily lives..[52]

Parents used toys and pastimes to instill practical and moral lessons.[53] For Magdalena Miller, from Knox, North Dakota, these lessons came through social activities and personal responsibilities. She remembers the joy of attending house parties and waltzing with her father, experiences that taught her grace, social etiquette, and the importance of community gatherings. She also recalls learning to care for her possessions. One year, when she was a young girl, she received a magnificent doll from her parents. However, her mother only allowed her to play with it on Sundays, reinforcing the value of patience, moderation, and respect for belongings. Thinking back on her childhood, Magdalena ap-

---

[50] Interview with Polly Berger.

[51] Interviews with Alice Buck, Drefs, Marvin Fiechtner, Jacob Helbling, Herman, August Heyne, Hartmann, Ost, and Weber.

[52] Interview with August Heyne.

[53] Interviews with Alice Buck, Haupt, Magdalena Miller, Norma Rost, Harold Schlenker, Mildred Wald, and Werth.

Image 6.3. Dakota Memories participant Marilyn Jeanette (Sauer) Bauman made this doll as a young girl, illustrating the creativity, resourcefulness, and personal pride central to German-Russian childhoods. *Photograph by William H. Clark, 2007; Marilyn Jeanette (Sauer) Bauman Collection, GRHC.*

preciates her mother's strictness, as that doll survived a number of years in the Miller home. Much like dancing taught her poise and self-control, this lesson in restraint deepened her appreciation for what she had, shaping how she valued both experiences and possessions.[54]

Through their pastime activities children also learned how to problem-solve or to reinvent their surroundings. Dakota Memories narrators com-

[54] Interview with Magdalena Miller.

plain about their lack of manufactured toys and sporting equipment during their oral histories. They also acknowledge that this never stopped them from having fun.[55] In the words of Clarence Bauman of Java, South Dakota, "[I]nnovate[d] as young kids are," they made their own fun despite the lack of store-bought toys. Clarence remembers creating his own fun at school, especially during the winter months. Their school was located near a hill, and so in the winter, children in Clarence's country school really enjoyed transforming that hill into a sledding paradise. According to Clarence, his family could not afford a store-bought sled. Thus, he and his younger brother, Harold, decided to make their own sled. They found a pine board, which he estimates at approximately six feet long by one foot wide. They polished up this board, and in no time at all, they had a makeshift sled. Clarence remembers having fun taking their sled up the hill and riding it down. Usually, they had three people on it at one time. Clarence's memory reflects a common experience among German-Russian children; resourcefulness was a necessity, but also a source of creativity and joy. Through these acts of ingenuity, children not only entertained themselves but also developed problem-solving skills that would serve them throughout their lives.[56]

Similar to Clarence, Wilfred Boechler also remembers having limited access to manufactured toys. When asked about his childhood toys, Wilfred reminisces about carving and whittling his own figurines. He remembers learning to carve from his older brothers, Frank, Kasper, Gregory, Charles, and Anton. They made coyotes, jackrabbits, foxes, and camels. Wilfred declares his brother Kasper could draw any animal, and his other brothers, himself included, carved Kasper's sketches out of wood. Wilfred points out that they never had any special carving instruments, just a standard pocketknife. Sometimes this was dangerous. On one occasion, when Wilfred was carving his figurines, his hand slipped. He nearly cut off his finger, which dangled from his hand. Wilfred remembers pushing his finger back into place and heading for the kitchen. He found some ointment in a cupboard, and wrapped up his finger in cloth. Even though he never saw a doctor, the "darn thing grew [back] together," Wilfred exclaims.[57]

---

[55] Interviews with Clarence Bauman, Wilfred Boechler, Polly Berger, Alice Buck, Graf, Klundt, Norma Rost, Mildred Wald, and Weber.

[56] Interview with Clarence Bauman.

[57] Interview with Wilfred Boechler.

Inevitably, children left to create their own fun found themselves in some mischief from time to time.[58] When Marvin L. Hartmann was a young boy in Demont, South Dakota, he participated in a few questionable activities. In particular, he recalls transforming his mother's Montgomery Ward or Sears Roebuck catalog into a target, inside the house. On this one occasion, Marvin recalls placing a chair in front of the main door, positioning the catalog on the chair, and shooting at his target using his BB gun. Since at this point Marvin was not a perfect shot, he did occasionally miss his target, permanently wedging BBs throughout his mother's kitchen. His mother was not exactly pleased with his ingenuity.[59]

Sometimes parents or older siblings assisted children in the fun-making.[60] For instance, Polly Berger remembers her father taking an active role in helping his children reinvent existing items and places. "We manufactured fun," exclaims Polly; "I was never bored." She shares that her father, Michael Anton, enjoyed helping his children have fun. In fact, Polly remembers him inventing a merry-go-round on the Wock family farm. He used an old buggy wheel that still had its axle attached to construct this merry-go-round. She points out that one of the wheels had fallen off, making it useless for transportation. Therefore, her father dug a hole and stuck the axle into the ground, positioning the wheel approximately two to three feet above the ground. This invention was the perfect merry-go-round, Polly declares. It entertained three to four children at a time. They simply sat on the wheel and used their feet to spin it around in circles. Thinking back on this makeshift merry-go-round, Polly cannot help but smile. She has fond memories of watching her sister's long, beautiful hair flying in the wind and of getting extremely dizzy, watching her siblings go around in circles endlessly.[61]

Many of the pastimes that entertained German-Russian children also helped them form their identity as German-Russians in the United States and Canada. Only a few narrators have memories of playing distinctly German-Russian games or playing with distinctly German-Russian toys.[62] Games such as

---

[58] Interviews with Clarence Bauman, George Berger, Boschee, Marvin Fiechtner, Hartmann, Ost, Norma Rost, Harold Schlenker, Schmidt, and Mildred Wald.

[59] Interview with Hartmann.

[60] Interview with Polly Berger, Clarence Bauman, Wilfred Boechler, Alice Buck, Drefs, Jacob Helbling, Mildred Wald, and Weber.

[61] Interview with Polly Berger.

[62] Interviews with Wilfred Boechler, Gab, August Heyne, Michael Jahner, Reinhold Klein, Norma Rost, Schaffer, Mildred Wald, and Weber.

Image 6.4. The merry-go-round that Pauline "Polly" (Wock) Berger's father devised out of a wagon wheel, showcasing the ingenuity and love that turned farm scraps into joyful play for children. Riding the wheel are Polly's brothers, Wallace Joseph and Michael Anton Jr., *circa* 1940. *Photograph from the Pauline "Polly" (Wock) Barger Collection, GRHC.*

Fahren einander aus dem Land (Driving Each Other Out of the Land), *Belznickel* (a holiday game associated with German-Russian Christmas traditions), and card games rooted in German-speaking communities were among the few pastimes that directly connected children to their ancestry. Likewise, some children played with handcrafted wooden toys, cloth dolls, and other homemade items that reflected the frugality and craftsmanship of their families.

Born near Kintyre, North Dakota, Mary (Mitzel) Weber remembers playing a German card game with her father, Joseph: *Fahren einander aus dem Land* (*Driving each other out of the Land*). While Mary cannot recall the rules of this game, she does share her memories of a funny incident when her father pretended to get overly upset by the hand he was dealt. Apparently frustrated, Joseph threw his cards into the stove—or so Mary thought. Wanting to mimic her father, Mary, who deeply admired him, followed suit, and tossed her own cards into the fire. Stunned, her father looked at her and said, "That was stupid!" He then revealed that he had only pretended to throw his cards away. Mary, still a young girl at the time, had fallen for his playful trick. With their game now in ruins, they found some paper and simply made their own cards, turning the mishap into yet another moment of fun and ingenuity.[63]

Learning to play American or Canadian games largely took place at school or church. Baseball, for instance, is an American pastime through and through. Mildred Wald remembers learning to play baseball from her older siblings, whom she believes learned the game from their father. Thinking back on her childhood, Mildred remembers they always had access to worn and used baseball equipment, such as gloves, bats, and balls. She enjoyed running around outside, despite her lack of talent. Sometimes, the whole family even went to Hague to watch the town team play on Sunday afternoons. Mildred says the best game to watch was the Hague versus Zeeland game, as they were town rivals.[64]

While many Canadian narrators reminisce about playing ball at an early age, more discuss their memories of hockey, a distinctly Canadian sport.[65] For generations, hockey has been a central part of Canadian culture, played on frozen ponds, in small-town rinks, and at the professional level—making it a time-honored tradition that connected communities and shaped childhood experiences. Hockey was Wilfred Boechler's favorite pastime as a young lad. He remembers when he first learned about the sport. He and his brother were listening to the radio when a live broadcast of a hockey game began. At first, they had no idea what they were hearing—just a lot of shouting and excitement—but they soon discovered it was a hockey game. Until that moment, they had never encountered the sport, a reminder that for some rural German-Russian children, cultural traditions like hockey were learned rather than inherited.

Shortly after hearing the broadcast, the Boechlers attended an amateur game at the town rink, where Wilfred finally saw hockey firsthand and learned its rules. Soon after, the Boechler boys took to the ice themselves, adapting their environment to fit their new pastime. They transformed shoes into skates, ponds into rinks, "horse turds" into pucks, and tree branches into sticks. This act of ingenuity—reshaping their surroundings to participate in a sport they

[63] Interview with Mary Weber.

[64] Interview with Mildred Wald.

[65] Interviews with Wilfred Boechler and Jahner.

had only just discovered—illustrates the adaptability of German-Russian children in Canada as they blended their heritage with the traditions of their new homeland.[66]

Educational pastimes shaped the identity of the second- and third-generation German-Russians. The lessons attached to playing hockey or baseball, transforming run-down vehicles into playground equipment, reinventing discarded items as toys, or making something from scratch were profound. While enjoying their childhood, these children learned about their gender roles, occupational requirements, imagination and creativity, and national identity. Moreover, these educational pastimes contributed to the emergence of a revised German-Russia story. The majority of narrators treasure these pastimes, as they altered the German-Russian story by providing the descendants with new, positive experiences and memories that overshadowed the pessimistic narrative.[67]

During the so-called "century of children," societies began to recognize a new sub-section of childhood: adolescence.[68] This stage in development was liminal, meaning the youth no longer belonged in the category of children, but were not yet adults. This distinction is important in discussions of play because, unlike younger children, adolescents do not engage in or describe their pastimes as "play."[69] After the seventh or eighth grade (thirteen to fourteen years of age), children no longer want to play make-believe with dolls and trucks, or play games such as Fox and Geese or Pump, Pump, Pull Away. To them, these activities are too childish and too simple. Instead, their leisure time becomes structured around activities, competitions, and hobbies—such as sports, organized social events, or skill-based tasks—that reflected their growing maturity. Because of this shift, youth do not neatly fit into the earlier categories of enjoyable and educational play, even though the pastimes they retained have

---

[66] Interview with Wilfred Boechler.

[67] Interviews with Clarence Bauman, Polly Berger, Wilfred Boechler, Alice Buck, Gab, Graf, Hartmann, Jacob Helbling, Herman, August Heyne, Hoff, Jahner, Reinhold Klein, Klundt, Magdalena Miller, Ost, Norma Rost, Schmidt, James Vetter, and Weber.

[68] West, *Growing Up in Twentieth-Century America*, 2.

[69] For more information on the idea of liminality, read Victor Turner, "Liminality and Commuitas," in *The Ritual Process: Structure and Anti-Structure* (Ithaca, NY: Cornell University Press, 1969), 94–130.

elements of both.[70] Unlike childhood games and toys, which were often framed as play, adolescent activities were described in ways that aligned more with responsibility, skill-building, or social engagement. This distinction is critical in understanding how pastimes evolved as German-Russian children matured: while they still engaged in forms of recreation, they no longer viewed their activities in the same way they did as younger children. The transition from childhood to adolescence marked a turning point in how play was perceived, a topic that will be explored in greater depth in the coming-of-age chapter.

As German-Russian children grew older, their pastimes reflected their transition from childhood to adolescence. Rather than engaging in games and imaginative play, they sought out social activities that reinforced their growing independence. Courting and dating became a central part of youth culture, as did social outings beyond the farm.[71] Clarence Bauman recalls how older children at school would sneak off into the woods during recess, enlisting younger students as lookouts while they "had a little excitement of their own."[72]

For many, weekend dances, movies, and parties became the primary forms of recreation. Attending Saturday night barn dances was a common activity among German-Russian youth, sometimes even for those technically too young to drink. Adam Boschee remembers meeting new people at these events, and, at times, getting into mischief. "Even though we weren't of age," he recalls, "there was always beer in the alley or in somebody's car." The dances and gatherings provided a space for socialization, self-expression, and connection with peers going through the same transition to adulthood.[73]

Athletics became a defining feature of adolescence for many German-Russian youth. Teenagers like Mary Sellinger (*circa* 1940s) participated in structured team sports such as softball, reflecting a shift from informal childhood games to organized competition. Unlike childhood play, which was often spontaneous and imaginative, sports introduced elements of discipline, teamwork, and performance. School-based athletics and community teams reinforced values of perseverance and commitment, while also providing opportunities

[70] Interviews with Clarence Bauman, Fercho, Gab, Jacob Helbling, August Heyne, Ost, Mildred Wald, and Schaffer.

[71] Interviews with Hilda (Lindemann) Backfish, Clarence Bauman, Delores (Kuhn) Becker, Polly Berger, George Berger, Alice Buck, Dewald, Florence (Mitzel) Engelhart, Henrietta (Gall) Fiechtner, and Erwin Gieszler.

[72] Interview with Clarence Bauman.

[73] Interview with Boschee.

Image 6.5. Dakota Memories interviewee Mary Sellinger (back row, first from the left) played softball with the Vics for a number of years during her youth, *circa* 1940s. This image highlights how sports fostered community pride, social bonds, and new opportunities for young German-Russian women. *Photograph from the Mary Sellinger Collection, GRHC.*

for social bonding. Whether playing softball, hockey, or baseball, adolescents engaged in pastimes that mirrored their increasing responsibilities and their desire to integrate into broader social circles.[74]

This shift in recreation—from unstructured play to more goal-oriented activities—reflect the larger transition from childhood to adulthood. As German-Russian youth matured, their pastimes became less about make-believe and more about self-expression, independence, and preparation for adult life. Activities such as sports, dating, and attending social events allowed adolescents to navigate their place within their communities. While they may not have spoken about these pastimes in the same nostalgic way they recalled childhood play, they were nonetheless formative experiences, shaping how the German-Russians saw themselves and their evolving identities. The coming-of-age chapter will explore this shift further, examining how German-Russian youth balanced their ethnic heritage with the cultural expectations of adulthood.

[74] Interviews with Mary Sellinger and Wilfred Boechler.

According to existing childhood and play historiography, play is essential to child development. Scholars such as West, Riney-Kehrberg, and Holt have already established the importance of play in rural childhood.[75] In their respective works, West, Riney-Kehrberg, and Holt discuss the role of toys, games, and activities in shaping the identity of children. For instance, in *Children of the Western Plains*, Holt acknowledges children created homemade toys for a variety of reasons, such as "the sentimental personal touch of making something for a loved one; the skills children learned making an item; the lack of money to purchase manufactured goods; and isolation from trading centers."[76] She continues by recognizing the ingenuity of rural children, as they transformed their parents' trash into their treasured toys, such as the Miller girls and their makeshift pie pans.[77]

Historians of child play, such as Cross and Chudacoff, also provide insight into the role of pastimes in child development and identity formation. In their foundational works, *Kids' Stuff* and *Children at Play*, respectively, they examine the role of play with toys, games, and activities in child development. Cross focuses on the evolution of the American toy industry, and its impact on the changing concept of childhood. He concludes that enjoyable and educational play have a role in the identity formation of American children. In fact, Cross aptly titles one of his chapters "Building Blocks of Character." He also believes that play was not "the opposite of work but another form of it."[78] It provided children with opportunities for a variety of lessons and experiences in all realms of life, including gender roles, occupational training, problem-solving skills, and national identity. In the words of Cross toys and pastimes "were to train children to be adults through imaginative play."[79]

This new research on pastimes of German-Russian children reaffirms and expands the conclusions in the existing historiography on child play. As Dakota

---

[75] For more information, see West, *Growing up with the Country*; Riney-Kehrberg, *Childhood on the Farm*; Holt, *Children on the Western Plains.*

[76] Holt, *Children of the Western Plains*, 130.

[77] Holt, *Children of the Western Plains*, 130; Interviews with Alice Buck and Klundt.

[78] Cross, *Kids' Stuff*, 127.

[79] Cross, *Kids' Stuff*, 127; Cross, "Gendered Futures/Gendered Fantasies: Toys as Representation of Changing Childhoods," *The America Journal of Semiotics* 12 (1998); Cross, "Toys and Time: Playthings and Parents' Attitudes Toward Change in early 20th-century America," *Time & Society* 7, no. 1 (1998); Chudacoff, *Children at Play.*

Memories narrators attest, they played a variety of games and activities during their childhood. Through these activities, German-Russian children had fun, while also learning about their gender roles, ethnic traditions, regional customs, and national identity. By examining play within a specific ethnic group—German-Russians—this research provides new insight into how childhood recreation contributed to the process of Americanization and Canadianization.

Rather than experiencing assimilation (fully adopting the dominant culture) or acculturation (adapting while retaining distinct cultural elements), Dakota Memories narrators reveal cultural diffusion—a process of selective borrowing and blending—occurred through their pastimes. They did not abandon their German-Russian heritage, nor did they passively absorb mainstream North American culture. Instead, they actively participated in new recreational activities, such as baseball and hockey, while maintaining elements of their ethnic identity within these spaces. By embracing new pastimes on their own terms, second- and third-generation German-Russians fused aspects of their traditional culture with their emerging national identity, demonstrating that play was not just a leisure activity but a crucial site of cultural negotiation and identity formation.

Moreover, by using oral histories to understand the role of play in child development and identity formation, this research reveals the changing narrative of German-Russians. The first-generation immigrants, having endured displacement, persecution, and the struggles of resettlement, framed their collective memory around survival rather than leisure. Their story emphasized hardship—fleeing oppressive regimes, facing discrimination in their new homeland, and enduring the physical and emotional toll of life on the northern plains. For them, childhood was often remembered as a time of sacrifice and work, not of carefree play. Initially, Dakota Memories narrators try to sidestep any discussion of fun or enjoyment in their oral histories. As the inheritors of this narrative, second- and third-generation Germans from Russia are deeply aware that stories of play and pastimes challenge the established collective memory of hardship. They are determined to maintain and reaffirm their ethnic group's identity as one of resilience, sacrifice, and perseverance. Thus, when recalling their childhoods, they instinctively emphasize labor, duty, and discipline, downplaying moments of joy.

Yet, as Dakota Memories oral historians discovered, with patience and careful questioning, narrators begin to recall a wealth of memories centered around enjoyable and educational play. Once surfaced, these memories—rather than those of hardship—are the ones narrators truly treasure, revealing a fuller, more nuanced picture of growing up German-Russian in North America.

Many German-Russian children carried fond memories of pastimes that had been passed down through generations. Leona (Kuhn) Hoff, from Dunn County, North Dakota, recalls *Babes in the Woods*, a haunting melody that had been sung to children for centuries. A traditional ballad, the song dates back to at least the seventeenth century and tells the sorrowful tale of two abandoned children who perish in the forest, later covered by leaves brought by robins. Over time, it became widely known as a lullaby and children's song,

My dear do you know, how long time ago,
Two poor little children, whose names I don't know,
Were stolen away, on a fine summer's day,
And, left in the woods, as I heard people say.

And, when it was night, so sad was their plight,
The sun had went down, and the moon gave no light,
They sobbed and they sighed, and they bitterly cried,
Tell the poor little child laid down, and they died.

And, when they were dead, the robin so red,
Brought strawberry leaves, and over them spread,
And, all the day long, they sang them this song,
Poor babes in the woods, Poor babes in the woods.

For German-Russian children, pastimes were more than just entertainment—they were sources of comfort, cultural continuity, and emotional connection to their heritage. Even songs like *Babes in the Wood*, with their melancholy themes, became part of the fabric of childhood, passed down not just as cautionary tales but as reminders of the fragility of youth. While hardship remained central to their identity, their personal memories reveal a more layered experience—one where resilience coexisted with moments of joy and play.[80] In

[80] Interview with Hoff.

the end, these children were not just “poor babes” lost in hardship; they were part of a rich, evolving cultural legacy that blended endurance with play and hardship with resilience.[81]

[81] Interview with Hoff.

CHAPTER 7

Gott Ist Die Liebe: Religious Devotion

"Every night he prayed for us," Arthur R. Bender says with tears in his eyes and a tremor in his voice. "I can't tell you how it feels to know that your parents prayed for you every night." Arthur, a native of Wishek, North Dakota, believes this memory is so precious because his father prayed for family members unbeknownst to them. At least, Arthur was unaware at first of his father's prayers. Then, one warm, summer evening in the 1930s, a curious Arthur decided to figure out why his father, Fred Karl, went to bed before the sun went down, while the Bender boys prepared the farm equipment for the next day. Quietly listening outside his father's room, Arthur heard him say, in German, "Be with Walter, and bless him." His father continued this prayer replacing the name each time, until he had prayed for each of his children. This left Arthur with an indescribable feeling. Pulling out a handkerchief to dry his eyes, he must take a moment to regain his composure before continuing with his oral history.[1]

Narrators of the Dakota Memories project share many of Arthur's sentiments regarding religion. They insist religion was an integral part of their childhood, and as a result, it shaped their identity. During their oral histories, German-Russians of the second and third generations spend time reminiscing about church services and programs, such as baptisms, first communions, confirmations, weddings, and funerals. Although narrators do not outright articulate it, this conservative institution and its activities helped their ethnic group maintain the German-Russian identity across borders and generations. Regardless of religious denominations, Protestant or Catholic, churches served as a safe place for

[1] Interview with Arthur R. Bender.

people to absorb change. This social institution allowed Germans from Russia to blend their ethnic identity with their new culture, thus diffusing their national identity with their ethnic identity. In essence, through religion, Germans from Russia on the northern plains refashioned the German-Russian story.

This commitment to religion is rooted in their ancestral heritage. Long before Germans from Russia came to the northern plains, these people were devout. Timothy J. Kloberdanz, a scholar of German-Russian heritage, insists, "[T]he role of religion in the [Russian] colonies was an all pervasive one."[2] This immigrant group intertwined religion in all aspects of life—political, social, and economic. "Religion," as Kloberdanz says, "was a source of comfort and came to occupy a prominent place in their lives."[3] He also recognizes that "religious fever waned at times, but it always remained a vital part of Black Sea German culture."[4] Kloberdanz concludes that the church, in fact, was "a primary vehicle for ethnic expression and community cohesion" for these immigrants.[5]

One reason Germans in Russia opted to immigrate to Russia was to maintain their religious freedom. In *The Uprooted*, a work about ethnic immigrants, American historian Oscar Handlin writes, "[T]he immigrants began with the determination that their emigration would not destroy the ties that bound them to the church."[6] Germans from Russia, for example, immigrated in order to maintain their religious practices. Thus, in Black Sea German communities throughout the northern plains, the first social institution constructed was the church. According to Kloberdanz, German-Russians "realized that such structures were essential not only to their spiritual needs but to their social ones as well."[7]

Although religion divided Black Sea Germans into distinct groups, its role in children's lives was essentially the same.[8] German-Russian descendants remember their grandparents teaching them the value of religion and their parents enforcing religious devotion. They also recall the socialization religion

[2] Timothy J. Kloberdanz, "Volksdeutsche: The Eastern European Germans," 125.
[3] Kloberdanz, "Volksdeutsche," 125.
[4] Kloberdanz, "Volksdeutsche," 125.
[5] Kloberdanz, "Volksdeutsche," 140.
[6] Oscar Handlin, *The Uprooted*, 2nd ed. (Boston: Little, Brown and Company, 1973), 111.
[7] Kloberdanz, "Volksdeutsche," 140.
[8] Kloberdanz, "Volksdeutsche," 140.

Image 7.1. Wilhelmina Bast and Dakota Memories participant Mary Sellinger on the day of their first holy communion, *circa* 1930. This event marked a major rite of passage, representing spiritual growth and community pride for German-Russian families. *Photograph from the Mary and Agatha "Peggy" Sellinger Collection, GRHC.*

provided through services and programs.[9] While church services and activities could sometimes feel tedious, faith remained deeply embedded in daily life—woven into morning prayers, mealtime blessings, evening rituals, and the moral lessons that shaped their understanding of right and wrong.

Since Sunday church services are the linchpin of religious life, they played a prominent role in the lives of German-Russian children. According to anthropologist Sonya Salamon, "[F]armers are churchgoers, and religion imposes a rhythm to their lives: six workdays followed by the Sabbath as a day of rest."[10] The same structured routine extended to their children, whose weeks often revolved around Sunday services. For many Dakota Memories narrators, church was more than a place of worship—it was a social hub where they gathered with their ethnic community, reinforcing cultural ties and lifelong friendships.

[9] Arends, *The Central Dakota Germans,* 107–10.

[10] Sonya Salamon, *Prairie Patrimony: Family, Farming & Community in the Midwest* (Chapel Hill: The University of North Carolina Press, 1992), 47.

Beyond weekly services, churches also served as the setting for life's most significant milestones. Historian Jane Marie Pederson notes that these institutions were where families "honored important turning points in the life cycle of an individual and of a family."[11] Baptisms, first communions, confirmations, weddings, and funerals were not just religious sacraments but community events that reinforced ethnic identity and social cohesion. In this way, churches functioned as both spiritual and social institutions, shaping not only faith but also the communal bonds of German-Russian families.[12]

Germans from Russia on the northern plains religiously observed four or five life-cycle rituals in their churches. In *The Central Dakota Germans*, linguist Shirley Fischer Arends—herself of German-Russian descent—recognizes the importance of some markers in German-Russian communities. She insists each baptism, confirmation, wedding, or funeral was a time of community togetherness. She writes that for the Lutheran church, "baptism marks a child's formal acceptance into the congregation and is a special occasion not only for the family but for the community" as well.[13] Regardless of the denomination, baptisms, confirmations, weddings, and funerals, were special community events for all church members.

As social institutions, churches also sponsored a number of community activities or events. In *Between Memory and Reality*, Pederson found that churches in ethnic communities "became the center of the social life among the settlers."[14] Moreover, she says "through the years they would sponsor numerous revivals and organize a host of activities in the community ranging from ice cream socials to temperance groups."[15] Through these religious traditions, Pederson concludes, ethnic groups had "a means of understanding their experience and adapting to it."[16]

German-Russian churches were no exception. In her study of regional dialect variations, Arends discusses the important social role of churches in ethnic communities. She describes the celebrations of the fall harvest, Christmas, New Year's Day, and Easter to name a few.[17] Thus, German-Russian churches,

---

[11] Pederson, *Between Memory and Reality*, 136.

[12] Pederson, *Between Memory and Reality*, 129.

[13] Arends, *The Central Dakota Germans*, 112.

[14] Pederson, *Between Memory and Reality*, 129.

[15] Pederson, *Between Memory and Reality*, 136.

[16] Pederson, *Between Memory and Reality*, 124.

[17] Arends, *The Central Dakota Germans*, 119–31.

like the churches Pederson examines, "defined the community and acted as centers of family and community rituals and sociability, drawing together large numbers of people."[18]

In her study of memory and identity, Pederson astutely states that "besides the family, the institution [that] carried much of the responsibility for cultural transmission on the frontier and in rural areas was the church."[19] She continues, "[W]ithin churches, immigrants preserved their language, their values, and their social traditions."[20] For German-Russians on the northern plains, churches did indeed help them transmit their language, culture, and identity, as well as help them adapt to their new homeland. This adaptation enabled the continuation of the German-Russian identity in the United States and Canada. "A sense of peoplehood or ethnicity is deeply tied to a group's religion," Pederson writes.[21]

By providing ethnic groups with a central place to share their traditions of the past and the present, churches essentially provided the necessary elasticity for the resilience of ethnicity. In *Preserving the Family Farm,* historian Mary Neth writes, "[E]thnic communities wanted both to maintain their ethnic identities and to enable their children to be successful in their new country."[22] For that reason alone, ethnic groups neither casually abandoned their ethnic identity, nor resisted adapting to mainstream culture. Rather, as in the case of German-Russians, some ethnic groups went through a process of cultural diffusion, by blending their ethnic identity with their national identity. Neth states that it was no easy feat to "[maintain] cultural tradition while becoming 'American,'" or Canadian. She continues, writing that it "was difficult for communities whose sense of neighborhood was strongly linked to ethnicity."[23] Yet German-Russians remained distinct longer than many other ethnic groups on the northern plains because their churches provided a framework for this diffusion. Religious institutions reinforced cultural traditions through German-language services, faith-based festivals, and community rituals that allowed younger generations to merge aspects of mainstream culture with their heritage rather than abandoning it.

---

[18] Pederson, *Between Memory and Reality*, 130.
[19] Pederson, *Between Memory and Reality*, 116.
[20] Pederson, *Between Memory and Reality*, 117.
[21] Pederson, *Between Memory and Reality*, 116.
[22] Mary Neth, *Preserving the Family Farm: Women, Community, and the Foundations of Agribusiness in the Midwest, 1900–1940* (Baltimore: John Hopkins University Press, 1995), 84.
[23] Neth, *Preserving the Family Farm*, 84.

Participants of the Dakota Memories project reminisce about a variety of religious activities, like confirmation, Sunday School, and church programs. While each denomination has differing religious beliefs and practices, the role of the church in the lives of children was quite similar. In other words, Lutherans, Evangelicals, and Catholics believe their churches played a significant role in their childhoods, and as a result, their lives. Their memories of church center around common themes: preparation and travel, Sunday worship (including service, language, hymns, socializing, and food), life-cycle celebrations (such as baptisms, first communions, confirmations, weddings, and funerals), and church programs (like Midnight Mass, Easter Vigil, and revival meetings). Through these activities, German-Russian descendants adapted aspects of their religious traditions to fit their evolving lives in America and Canada, blending old-world customs with new-world realities. This ability to adjust while maintaining core traditions reinforced their ethnic identity, ensuring its continuity across generations.

For Dakota Memories narrators, Sunday worship marked either the beginning or the end of the week. Thus, narrators have fond memories of preparing for this big day. According to the narratives, families typically reserved all or part of the day on Saturday for church preparation. This preparation usually entailed weekly baths, meal preparation, and setting out clothes. Preparing for Sunday service also involved transportation. Since people lived miles from the nearest church, and the weather was unpredictable on the northern plains, getting to and from church could be challenging. In spite of this, German-Russians rarely skipped a Sunday service.[24]

Elsie (Just) Johnson shares her memories of Saturday preparation as a member of a Lutheran family. She remembers taking a bath only once a week on Saturday evenings. The twelve Just children (six of whom were born in Russia, two of whom were born in Iowa, and four of whom were born in Wisconsin), as well as their parents, took their baths every Saturday because church was the next morning. She reminisces about fetching the water from the pump house, heating the water on the stove, taking a bath, then dumping the dirty water outside, and repeating the process for every member of her family. She also remembers wearing clean flannel nightgowns on Saturday nights. While

[24] Interviews with Nicolas F. Axtmann, Arthur R. Bender, Anne (Frehlich) Hoffart, interview; Elsie (Just) Johnson, Malvin W. Miller, and Irene (Bender) Schauer.

the children bathed, Elsie's mother ironed these nightgowns, and placed them in the children's beds.[25]

Growing up in Danzig, North Dakota, in a Baptist family, Irene (Bender) Schauer (Arthur R. Bender's younger sister) also remembers all of the preparation that took place on the Bender farm before church service on Sunday. She recalls that although it was a special day, it started like any other day. Her parents, Fred Karl and Emelia (Miller), did not excuse their children from their daily chores. The Bender children still had to milk the cows and feed the chickens before getting ready for church. While the children completed their morning chores, their mother laid out their "Sunday clothes." Irene especially enjoyed Sunday mornings because her mother always used a curling iron heated by the kerosene stove to curl Irene's hair. Holding the curling iron from her childhood, Irene exclaims with joy, "You curled your hair a bit," the simple act still bringing back fond memories of the care and attention that made Sundays different from every other day.[26]

For many German-Russian families, however, the joy of Sunday mornings was accompanied by the challenge of getting to church, especially in the winter. Born in Tuscarora Township in Pierce County, North Dakota, Nicolas F. Axtmann remembers the difficulties his family encountered traveling to and from their Catholic church in the winter months. He says they never missed church, if they could help it. To demonstrate the difficult conditions, Nicholas shares his memory of traveling to Midnight Mass one year after a few days of massive snow accumulations. He remembers the snowbanks were unbelievably high. Determined not to let the snow deter them from their religious worship, all eight of the Axtmann children and their parents loaded into an enclosed sled, or in Nicolas's words the "bus," and headed for church. On their way there, they tipped the bus twice. The journey that evening was long and dangerous.[27]

Regardless of their denomination, German-Russian descendants who participated in Dakota Memories all recall the central role of Sunday worship in their childhoods. Narrators describe the experience in vivid detail, from the structure of the building and the segregated seating arrangements to the cadence of the service, the language spoken, and the role of church leaders. They remember the power of music—whether through hymns, choirs, or instru-

---

[25] Interview with Elsie Johnson.

[26] Interview with Irene Schauer.

[27] Interview with Axtmann.

ments—and the sense of community fostered through post-service traditions, where food and conversation were as much a part of the morning as the prayers themselves. These memories, rich with personal stories, reveal the depth of religion's influence on their upbringing.[28]

Physical descriptions of church buildings vary depending on whether or not narrators attended rural country churches or town churches. Differences in the décor denote the economic situation of regions, as well as religious affiliations of churches. Originally, these buildings were simple structures, with altars for worship and pews for the congregation. During the early 1900s, however, German-Russian communities throughout the northern plains were remodeling or building new churches. Narrators recall some of the most striking examples from this era, including Saints Peter and Paul Catholic Church in Strasburg, North Dakota, and Saint Michael's Catholic Church in Tramping Lake, Saskatchewan —both admired for their architectural beauty and enduring presence in their communities.[29]

Church members usually kept the buildings well maintained, as these were, after all, sacred places of worship.[30] Irene (Bender) Schauer insists that her family "went to church every Sunday." She remembers one of her older brothers usually headed to church at least one or two hours before the service in order to warm up the building in the winter, as it was not heated during the week. "It was usually still dark when he left," Irene points out. Like many German-Russian churches, services were often held in the morning, typically beginning around 9:00 or 10:00 a.m. and lasting an hour or more, depending on the denomination and the length of hymns, sermons, and prayers. Irene always expected the church to be warm upon arrival; however, it usually took longer than expected. "By the time it warmed up, it was actually noon, and we were headed home," she says, as the service had already ended.[31]

In their reminiscences of the church services and its layouts, Dakota Memories narrators mention the separation of gender and age that existed for seating

---

[28] Interviews with Hilda Melita (Lindemann) Backfish, Pauline "Polly" (Wock) Berger, Wilfred Jacob Boechler, James Engelhart, Erwin Gebhardt, Erwin Geiszler, Louis Helfrich, Hoffart, Jerger, Elsie Johnson, Edwin Kittler, Reinhold Klein, Merkel Jr., Malvin W. Miller, and Schauer.

[29] For more information, see James Coomber and Sheldon Green, *Magnificent Churches on the Prairie: A Story of Immigrant Priest, Builders and Homesteaders* (Fargo, ND: Institute for Regional Studies, NDSU, 1996).

[30] Interviews with Backfish, James Engelhart, Kittler, and Schauer.

[31] Interview with Schauer.

Image 7.2. A Reformed Church built in the 1880s in Greenway, South Dakota. Dakota Memories participant James Engelhardt purchased the church in 1929 and maintains it on his property near Ashley, North Dakota. This building reflects the deep pride and commitment to preserving faith and heritage among German-Russian families. *Photograph by Jennifer Wilkie on May 23, 2006; James and Florence (Matz) Engelhardt Collection, GRHC.*

in the church until the World War II era began.[32] In his *Plains Folk* chapter on North Dakota's "other Germans," anthropologist Timothy J. Kloberdanz writes, "among both Catholic and Protestants, even the seating pattern in the churches tended to follow old country custom. Females traditionally sat on the left, while the men and boys took their seats on the right."[33] According to Kloberdanz, the shift to sitting together as families resulted from the process of Americanization and Canadianization.[34]

Born into a Methodist family in McPherson County during the 1930s, William Adam Merkel Jr. is one German-Russian Baptist who reminisces about the separated seating in his church, where the men sat on the right and the women on the left. In his Evangelical church, as was the case in other churches in this region at the time, the members also segregated themselves by age, youngest in the front and oldest in the back. Occasionally, this could lead to problems, as parents might be a few pews removed from their children. Although William remembers women primarily taking care of children during Sunday worship, he recalls his father poking him from time to time for whispering during the sermon. William's chuckle indicates this may have happened more than once.[35]

Narrators of the Dakota Memories project also reminisce about Sunday worship services. Although a few narrators remember the services as long and drawn out, sometimes lasting two to three hours, most narrators have a deep appreciation for them. In their memories about Sunday worship narrators discuss the church leadership (pastor, minister, or priest), Sunday School, and language. These stories about Sunday church service seem to mesh with the accounts provided by German-Russian scholars, like Arends and Kloberdanz. For instance, in *The Central Dakota Germans*, Arends maintains that in German-Russian communities ministers typically served more than one congregation, children attended Sunday School, and in the mid-1940s to early 1950s congregations witnessed the official church language transition to English.[36]

---

[32] Interviews with Axtmann, Wilfred Boechler, Ervin Gebhardt, Helfrich, Hoffart, Jerger, Elsie Johnson, and Merkel Jr.

[33] Kloberdanz, "Volksdeutsche: The Eastern European Germans," 141.

[34] Kloberdanz, "Volksdeutsche: The Eastern European Germans," 141.

[35] Interview with Merkel Jr.

[36] Interviews with Axtmann, Backfish, Bender, Wilfred Boechler, James Engelhart, Ervin Gebhardt, Gieszler, Helfrich, Jerger, Kittler, Merkel Jr., Malvin Miller, Schauer, interview; Arends, *The Central Dakota Germans*, 107; Kloberdanz, "Volksdeutche The Eastern European Germans," 141.

Erwin Geiszler, of Logan County, North Dakota, shares his memories of how a traveling pastor at Emmanuel Lutheran Church was responsible for serving four other regional churches in the 1920s. As a result, Erwin says his congregation only had a minister once a month. On the other Sundays, Erwin's father, John, led the services. According to Erwin, his father had this responsibility because he was one of the few farmers in the region with an education beyond grade school. Although John never mentioned it to his son, Erwin later learned from his paternal aunt that his father had received a high school education and had briefly attended seminary school. The story maintains that John had a heart condition, and his parents (Erwin's grandparents) believed that sending him to school would provide a safer and more sustainable future than the physical demands of farm labor. Because of this background, John was one of the only church members who could read and write in both German and English, making him a natural choice to lead services in the absence of a pastor.[37]

Dakota Memories narrators, like Arthur Bender, have vivid memories of the language transition to English. Some remember a gradual change, while others remember a rapid change. Depending on their religious affiliation, the shift was from the German language to English for Protestants or, for Catholics, Latin to English. Regardless, this change affected the first-generation Germans from Russia, as well as their descendants.[38] "One of the most difficult adjustments that Germans-Russian congregations had to make," according to Kloberdanz, "was the transition from German to English in their religious services."[39] Growing up in the south-central region of North Dakota, Arthur remembers this issue divided his family. Seeing as the Bender children were born between 1915 and 1922, they all learned English in grade school, so they welcomed the transition to English. Arthur remembers Sunday School being hard for the Bender children because the lessons were all in German, but they could neither read nor write in their native tongue. Their parents always had to help them prepare for Sunday School on Saturday evenings. However, Arthur's parents

[37] Interview with Geiszler.

[38] Interviews with Axtmann, Backfish, Bender, James Engelhart, Hoffart, Jerger, Elsie Johnson, Kittler, Reinhold Klein, Leier, Merkel Jr., Dorothy (Grasser) Miller, Malvin Miller, Schauer, and Gorden R. Werth.

[39] Kloberdanz, "Volksdeutsche: The Eastern European Germans," 141.

resented the language transition. Arthur insists his father continued to attend church every Sunday, but his father no longer understood the sermons after the 1950s when their congregation voted for a rapid transition to English only services. He remembers his father was quite disappointed with all the Bender children for supporting this transition.[40]

For Dakota Memories narrators one of the highlights of church was the musical performances during Sunday services. They thoroughly enjoyed the choirs, organists, and hymns from their childhoods, and for many this love of music carried into adulthood. In fact, several narrators remained active in church choirs or German choirs at the time of their interviews, still singing the hymns from their youth. Some even volunteered to sing during their oral histories, and on occasion played instruments, most commonly the organ though some churches also included other instruments such as accordions or brass ensembles.[41] Reinhold R. Klein, a member of the United Church of Christ and a retired clergyman, has particularly fond memories of church music, which was likely influenced by his father's role as the church choir director. "Church was very important," Reinhold emphasizes, recalling how hymns, like *Gott ist die Liebe* (God is Love) or *Großer Gott, wir loben dich* (Great God we Praise you) were central to every Sunday service. During his oral history, Reinhold even sang "Love Divine: All Loves Excelling," a hymn originally written by Charles Wesley in 1747 and commonly sung in both German and English congregations:

Love Divine, all love excelling,
joy of heaven, to earth come down;
fix in us thy humble dwelling;
all thy faithful mercies crown!
Jesus thou art all compassion,
pure, unbounded love thou art;
visit us with thy salvation;
enter every trembling heart.[42]

This hymn, widely recognized across Christian denominations, was a staple of worship services and remains a cherished part of religious tradition for many German-Russian descendants.

---

[40] Interview with Bender.

[41] Interviews with Polly Berger, Helfrich, Reinhold Klein, and Schauer.

[42] Interview with Reinhold Klein.

German-Russian memory-makers of the second and third generations also found the social aspect of Sundays memorable. Narrators reminisce about chatting with neighbors before church services, gathering with extended family after worship, and enjoying hearty Sunday dinners featuring chicken noodle soup, roasted chicken, fried chicken, or canned chicken. Afternoons and early evenings were spent outdoors, with children playing ball with friends while adults continued their conversations.[43] For Malvin W. Miller , a Lutheran from Zap, North Dakota, visiting—whether stopping by a neighbor's farm, welcoming relatives into their home, or joining friends for an afternoon meal—was simply part of everyone's Sunday plans. Fried or roasted chicken was always on the table and baseball games with neighbors were a weekly ritual. As Malvin fondly remembers, "It was a great day."[44]

The church played a significant role in recognizing important life cycle events.[45] In *Between Memory and Identity*, Pederson recognizes these "life markers" as essential rites of passage that shaped both individual lives and communal traditions. Baptism, the first of these milestones, varied by denomination. In Catholic, Lutheran, and some Protestant congregations, it was performed in infancy to welcome the child into the faith, while in Baptist, Evangelical, and Seventh-day Adventist traditions, it was often reserved for older children or adults who could make a personal declaration of faith. Confirmations, particularly significant in Protestant traditions, marked a young person's transition into adolescence and their deeper commitment to religious beliefs. Catholic Germans from Russia also observed first communion, which signified a child's formal participation in the Eucharist and their growing religious devotion. Weddings were central across all denominations, not only as religious ceremonies but also as community gatherings that reinforced family and social ties. Funerals, the final life marker, honored the passing of loved ones and provided a communal space for mourning and remembrance, often blending faith traditions with ethnic customs. "Not only did these rituals acknowledge the significant events in life," says Pederson, "but they were also an important component in the organization of social life for a community."[46] German-Russian

[43] Interviews with Polly Berger, Helfrich, Hoffart, Jerger, Elsie Johnson, Leier, and Malvin Miller.
[44] Interview with Malvin Miller.
[45] Interviews with Axtmann, Backfish, Polly Berger, Wilfred Boechler, James Engelhart, Geiszler, Helfrich, Hoffart, Jerger, Elsie Johnson, Kittler, Leier, Dorothy (Grasser) Miller, and Malvin Miller Schauer.
[46] Pederson, *Between Memory and Reality*, 137.

descendants recall how their churches used these ceremonies to bring communities together, strengthening faith and cultural identity.[47] These events, deeply rooted in tradition yet adapted to their new *heimat* (homeland), reinforced the German-Russian ethnic identity and ensured the endurance of their religious and cultural heritage across generations.

In many German-Russian churches, baptisms took place shortly after birth, thus narrators have no memory of their own baptisms, though some share stories of family and community members. Regardless of personal memory, as these memory-makers substantiate, baptisms were formal celebrations, involving both immediate and extended families, as well as the larger religious community. This ritual marked not only a child's entry into the faith but also reinforced communal bonds, as families gathered for the occasion. While baptism was a significant religious milestone, its occurrence in infancy meant it did not feature prominently in the personal childhood memories of most narrators. However, its importance was still deeply understood, as it signified the first step in a lifelong connection to both faith and ethnic identity.[48]

Although Elsie Johnson has no memory of her own baptism, she shares a family story about how the pastor had scheduled two baptisms on the same day, Elsie's and another girl's. In the spirit of neighborliness, Elsie's mother, Rosalie (Jescke), sewed the baptismal gowns for both infants. Elsie also recalls that the other child was baptized first, noting that her family had been members of the congregation longer than the Justs. Whether this order reflected intentional hierarchy or was simply a matter of timing is unclear, but the experience suggests the presence of social dynamics within the church, or at least how it is remembered by Elsie. Yet, Elsie also emphasizes that the occasion was a community celebration, honoring the lives of two new church members and reinforcing the bonds between families in the congregation.[49]

Unlike baptism, the ritual of First Holy Communion was remembered more from the childhoods of Dakota Memories narrators. For Catholic narrators, in particular, this event marked a pivotal transition in their lives. Since it typically took place in early childhood, around the age of five or six, narrators do not spend much time reminiscing about it, as their memories are usually

---

[47] Interviews with Axtmann, Backfish, Polly Berger, Wilfred Boechler, James Engelhart, Geiszler, Helfrich, Hoffart, Emmanuel Gene Jerger, Elsie Johnson, Kittler, Leier, Merkel Jr., Dorothy Miller, Malvin W. Miller, and Schauer.

[48] Interviews with Ervin Gebhardt, Herbert Herman, Elsie Johnson, Thurn, and Sperling.

[49] Interview with Elsie Johnson.

Tauf-Schein.

Wahrlich, ich sage euch, wer das Reich Gottes nicht empfähet als ein Kindlein, der wird nicht hineinkommen. Mark. 10:15

Wer da glaubet und getauft wird, der wird selig werden. Mark. 16:16

Elmar Donald

geb. den 10 ten Mai 1926 in Lehr, Logan Co. N. Dak.

Sohn des Herrn David Sperling

und dessen Ehefrau Sophia, geb. Essig

ist am 11 ten Juli 1926 in der Ev. Luth. Dreieinigkeits Kirche

im Namen des Dreieinigen Gottes getauft worden.

Taufpaten waren: Henry Essig

Clara Koenig

Walter A. Jemke, Ev. Luth. Pastor

Lasset die Kindlein zu mir kommen und wehret ihnen nicht; denn solcher ist das Reich Gottes. Luk. 18:16

Image 7.3. Elmer D. Sperling's baptismal certificate, 1926. This document reflects the deep spiritual commitment and the importance of formal faith milestones in German-Russian family life. *Document from the Elmer D. Sperling Collection, GRHC.*

vague. Therefore, when reminiscing about first communion, narrators avoid discussing the meaning of this ritual. Dakota Memories narrators typically discuss only the clothing, procession, and congregational attendance.[50]

Fran Schwan, of Rugby, North Dakota, vaguely remembers her First Holy Communion at Sacred Heart Catholic Church in Cando, North Dakota, in 1943, when she was barely six years old. "I really didn't understand it as well as I should've," exclaims Fran. She remembers her mother made her a magnificent white gown, yet they could not find a pair of white shoes to match. Fran suggests this shortage most likely correlated with the war efforts at the time. Thus, as Fran points out, she ended up wearing saddle shoes to the ceremony. Fran also remembers participating in a formal procession. The children formed a line outside the church before marching in together. While she admits this ceremony was a significant moment in her childhood, her memories of it have faded with the passing of time.[51]

Unlike baptism, which took place in infancy, or first communion, which occurred at a younger age, confirmation is a religious celebration that Dakota Memories narrators personally remember. Typically occurring between the ages of twelve and fourteen, confirmation marked a rite of passage in the lives of German-Russian immigrants and their descendants. At this stage, narrators were old enough to fully participate in the process and retain clear memories of their experiences.[52] Most congregations on the Northern Plains preferred to have their youth confirmed by the age of fourteen. Arends states, confirmation "was the child's acceptance into the faith and into the fellowship of the congregation as an adult with full rights."[53] Narrators of the Dakota Memories, in particularly Protestant narrators, remember participating in confirmation between twelve and fourteen years of age, and claim that following confirmation they believed they obtained more freedom. In reminiscing about this rite of passage, German-Russian memory-makers emphasize three specific elements, the classes, the ceremony, and the celebration.[54]

---

[50] Interviews with Boechler, Christina Rose Marie (Bast) Krismer, Schwan, Agatha "Peggy" Sellinger, and Mary Sellinger.

[51] Interview with Fran Schwan.

[52] Interviews with Axtmann, Backfish, Polly Berger, Wilfred Boechler, James Engelhart, Geiszler, Louis Helfrich, Anne Hoffart, Jerger, Jacob A. Helbling, Elsie Johnson, Leier, William Adam Merkel, Jr., Dorothy (Grasser) Miller, and Malvin Miller.

[53] Arends, *The Central Dakota Germans,* 113.

[54] Interviews with Axtmann, Backfish, Polly Berger, Wilfred Boechler, James Engelhart, Geiszler, Helfrich, Hoffart, Jerger, Kittler, Jacob Helbling, Elsie Johnson, Barbara Leier, Merkel Jr., Dorothy Miller; and Malvin Miller.

Image 7.4. LeRoy Buck was confirmed on the afternoon of Sunday, July 4, 1943, marking his transition to adult membership in the church and producing a source of deep family and community pride. Back row (left to right): LeRoy Buck, Ernest Buchholz, Loretta Wentz, Helen Ketterling, Orville Ketterling, and Alvin "Dynamite" Mayer. Front row (left to right): Lorraine Wentz, De Elda Bitterman, Reverend Gust C. Thiele, Elda Ketterling, and Jacque Bye. *Photograph from the LeRoy and Alva Ruth (Miller) Buck Collection, GRHC.*

For those who had confirmation classes they played a significant role in the lives of German-Russian children. Narrators have fond memories of their confirmation classes, which they typically took place over a couple weeks during the summer months. Thus, attending confirmation class was a way for German-Russian farm children to replace days of labor with days of education and socialization.[55] Edwin Kittler (1948, Hazen ND) grew up in a Lutheran family. He remembers confirmation classes lasting all day, five days per week, for three weeks. During the classes, Edwin says, children learned their catechisms, psalms, and bible history. He remembers they had a motivational pastor.[56]

[55] Interviews with Axtmann, Backfish, PollyBerger, Wilfred Berger, Helfrich, Jacob Helbling, and Hoffart.

[56] Interview with Kittler.

Children who attended a parochial school (or a public school located in a predominately Catholic region) typically had different experiences with confirmation classes. These children received their religious instruction in school, as part of their core curriculum.[57] Reared in a Roman Catholic family, Wilfred Jacob Boechler, from Allan, Saskatchewan, remembers receiving his religious instruction from the public school he attended. According to him, the religious classes always took place in the afternoon, so those who were not Catholic simply left early. Wilfred remembers about ten students outside the Catholic faith attending his school. The Allan School had four classrooms—one room and teacher for grades first through third, fourth through sixth, seventh through ninth, and tenth through twelfth—and four Catholic nuns as the teachers. While the nuns only reserved the last hour specifically for religious instruction, Wilfred insists that all instruction was rooted in religion.[58]

After children completed their confirmation classes, congregations recognized this accomplishment by holding the religious ceremony. These were community events and typically involved seven or more children. If there were not enough children one year, the congregation either had younger children go through the classes a year or two early, or they held older children back a year or two. The narrators remember dressing in beautiful gowns, participating in a procession, and presenting memorized scriptures for their congregations. For some, the pressure of reciting scripture publicly and the formality of the occasion made the experience nerve-wracking, while others cherished the sense of achievement, the attention from family and friends, and the recognition of their growing role in the church. While some recall feeling intimidated, others remember it as one of their best experiences.[59]

Hilda Melita (Lindemann) Backfish, a Lutheran from Golden Valley, North Dakota, remembers the grand ceremony for her confirmation class. She describes how all the girls wore crisp, new, white dresses, many of them handmade by mothers and grandmothers, with delicate lace and accents or carefully sewn pleats. The boys, dressed just as formally, wore new suits— some tailored, others purchased for the occasion—making one of the first times they had worn such fine clothing. In honor of the event, Hilda even had her first perm, a significant moment of preparation that made her feel both excited and grown-up. At

[57] Interviews with Wilfred Boechler and Louis Helfrich.

[58] Interview with Wilfred Boechler

[59] Interviews with Backfish, Malvin Miller, Wilfred Boechler, James Engelhart, Geiszler, Anne Hoffart, Jerger, Kittler, Elsie Johnson, Merkel Jr., and Dorothy Miller.

the end of the ceremony, she recalls her congregation presenting each child with a German hymnal, their names embossed in gold lettering on the cover. This hymnal meant the world to Hilda, a tangible symbol of her place in the faith and a treasured keepsake of a milestone that had been both solemn and joyous.[60]

German-Russian descendants also remember having a family or community celebration following this ceremony. Some Dakota Memories narrators recall only having a small family gathering, sometimes with just immediate family, while others remember more elaborate celebrations that included extended relatives, neighbors, and members of the congregation. These gatherings were not just about marking a religious milestone but about reinforcing the communal bonds that were central to German-Russian life, where faith and family were deeply intertwined. Narrators emphasize their parents recognized confirmation as a pivotal moment in their children's lives, symbolizing their growing responsibility within both the church and the broader community.[61]

As a member of a Lutheran family, Emmanuel Jerger, originally from Colorado, vividly remembers the commemoration his family had in honor of his confirmation. "It was a big celebration," he exclaims. Uncles, grandparents, and friends gathered to mark the occasion. The day was filled with food, laughter, and tradition as tables were laid with homemade dishes such as sausage, fresh bread, pickled vegetables, and sweet kuchen—a meal that reflected the heritage of those present. The gathering was a way of honoring both the individual and the collective, reinforcing the idea that faith was not just personal but something nurtured within the support of family and community. After eating, children took to the open fields for a game of ball, while adults lingered over coffee, sharing stories and laughter late into the afternoon. Whether small or large, these celebrations underscored how faith, family, and cultural tradition were inseparable in the German-Russian experience, making confirmation not just a personal rite of passage but a moment of collective significance.[62]

Another life marker commonly celebrated in the church was the union of marriage between two young churchgoers. Resembling German-Russian religious rituals, weddings typically involved the immediate and extended families and the church community. Through these celebrations, German-Russians practiced their traditions and even adopted new customs. According to the tes-

---

[60] Interview with Backfish.

[61] Interviews with Polly Berger, Jerger, and Merkel Jr.

[62] Interview with Jerger.

timonies of German-Russian narrators across the northern plains, weddings in the early 1900s were rooted in traditions, yet influenced by the process of Americanization or Canadianization. For instance, Dakota Memories narrators insist they married by choice and for love, unlike their parents and grandparents who were arranged in marriages.[63] Shirley Fischer Arends notes, "[M]atchmaking . . . remained a practice in the communities of the Germans from Russia until well into the 1920s," but narrators of the Dakota Memories have no personal recollections of matchmakers.[64] Instead, they reminisce about courting, love, engagements, and weddings—stories centered on friendships and the experience of finding romantic love rather than arrangements made by parents.[65]

Regardless of the denomination, Dakota Memories narrators recall wedding ceremonies taking place at church, followed by receptions usually at the family farm of either the bride or groom. Still, there are occasional accounts of weddings taking place in homes and receptions taking place at community halls or centers, and the length of the wedding celebrations did vary. Protestants celebrated for a few hours, whereas, Catholics celebrated at least twelve hours.[66] Edna Lu Ella (Weisz) Buck, a Methodist from Lehr, North Dakota, remembers her wedding on March 24, 1940, in great detail. "It was a beautiful day," she says, made even more special because it was the same day as her parents' anniversary. The ceremony took place at the Streeter Evangelical Church, where friends and family gathered to witness the occasion. Afterward, her parents hosted the reception at their home, a simple but heartfelt gathering. "There were no thrills," Edna says. They had sandwiches and cake for the guests, and "whoever wanted to come, came." The reception became a large communal affair, as neighbors, extended family, and friends arrived to celebrate. While some weddings were elaborate, hers reflected the openhearted hospitality of her family and community. The day was significant; a union shared with all who wished to be part of it.[67]

Pauline "Polly" (Wock) Berger, a Catholic from Stark County, North Dakota, remembers her wedding day as a lavish and joyous celebration. On October 20, 1947, after three years of courtship, Polly married Valentine Berger

---

[63] Interviews with Becker, Rosemary (Fisher) Berger, Wilfred Boechler, Henrietta (Gall) Fiechtner, and Helfrich.

[64] Arends, *The Central Dakota Germans*, 131.

[65] Interviews with Becker, Rosemary Berger, Wilfred Boechler, James Engelhart, Henrietta Fiechtner, and Helfrich.

[66] Interviews with Becker, Rosemary Berger, Polly Berger, Wilfred Boechler, Alice Buck, Edna Lu Ella (Weisz) Buck, LeRoy Buck, James Engelhart, Henrietta Fiechtner, Helfrich, and Herman.

[67] Interview with Edna Buck.

Image 7.5. Alice (Walter) and Valentine F. Brossart were married on October 23, 1950. This wedding marked an important cultural and family milestone, symbolizing continuity and community pride. *Photograph from the Valentine F. Brossart Collection, GRHC.*

in a ceremony that stretched from morning until night. "We were married in church," Polly says, "and the dinner was in a hall." After the religious ceremony, guests gathered for a lunch of chicken noodle soup and pie before the celebration continued with dancing. Supper featured ham, potato salad, and wedding cake, followed by even more dancing, and later in the evening, guests enjoyed punch and kuchen before the festivities finally came to an end. Polly recalls the day fondly, emphasizing its joyful atmosphere and the presence of family and friends who came together to mark the occasion. Weddings "were big festivities," Polly maintains.[68]

[68] Interview with Polly Berger.

Funerals were another life-cycle marker observed religiously by German-Russians, serving not only as moments of mourning but also as communal gatherings that reinforced social and cultural bonds. Somber occasions with little emphasis on celebration, they still maintained, as observed by Arends, "a sense of community."[69]

In *The Central Dakota Germans*, Arends notes that, "when a member of the Dakota German community dies, the church bell is immediately rung. It is rung three times, short rings for children, long rings for adults."[70] While narrators of the Dakota Memories project do not mention this practice, they recall other customs, such as as holding wakes in family homes, where the presence of the deceased created an eerie yet sacred atmosphere. Many narrators remember feeling unsettled by the tradition of keeping the body in the house for viewing, but they also describe it as an intimate way to say goodbye, a practice that strengthened family ties and honored the dead. Traditional German funeral songs filled these spaces, further shaping the experience of loss. Although narrators emphasize death as an ending, faith remained central to the mourning process, and for most, belief in Heaven provided comfort, reinforcing the idea that while life on earth concluded, the soul's journey continued.[71]

Paul Welder, from Strasburg, North Dakota, shares his childhood memories of funerals in the 1920s, insisting they felt much different back then than they do now. He remembers funerals from the days of his youth as overwhelming events filled with deep, uncontainable grief, with people sobbing uncontrollably, their cries echoing through the church. He recalls that some mourners even screamed in anguish. To a child, these raw expressions of sorrow made death feel even more unsettling, reinforcing the weight of loss in a way that was both frightening and profound. The solemn atmosphere was heightened by the haunting beauty of traditional German funeral songs, which Paul recalls vividly. Though he now understands that grief is a deeply personal and cultural expression, at the time, funerals seemed unbearably heavy—moments when sadness took hold of an entire community, leaving a lasting imprint on his memory.[72]

---

[69] Arends, *The Central Dakota Germans*, 137.

[70] Arends, *The Central Dakota Germans*, 137.

[71] Interviews with Marvin L. Hartmann, Herbert Herman, Jerger, Barbara Leier, Merkel Jr. and Schauer.

[72] Interview with Paul Welder.

Image 7.6. Funeral for Florence Bast, Christina Rose Marie's baby sister, 1949. This photograph captures the communal mourning and the importance of shared rituals in sustaining faith and supporting families during times of loss. *Photograph from the Christina Rose Marie (Bast) Krismer Collection, GRHC.*

William Adam Merkel Jr. and Emmanuel Jerger also described funerals as deeply emotional experiences. According to William, although he realized that "death was a natural part of life," the funeral services always made him feel uneasy. He remembers they commenced at church and ended with a prayer service at the cemetery with the lowering of the casket into a wooden box. This process, William recalls, usually involved a couple of men getting into the hole and nailing the wooden box closed. The entire community worked together at burying the casket. "That was hard work!" he concludes. Emmanuel (a devout Lutheran) shares a more personal memory. When Emmanuel was just seven years old, he watched his family lay to rest his older brother Jacob. For Emmanuel it was especially hard, since his family prepared the body at home. He remembers feeling a little relieved when the pallbearers finally arrived to transport the casket to the country church. His brother rests in a rural cemetery near Batesland, South Dakota.[73]

[73] Interviews with Merkel Jr. and Jerger.

Churches were a prominent social institution during the Dakota Memories project participants' childhoods. While narrators reminisce about Sunday worship and religious rituals, or life markers—baptisms, first communions, confirmations, weddings, and funerals—they also share memories of church-sponsored activities and celebrations. Again, these events brought together the larger community for common purposes. Dakota Memories narrators remember a host of events, including revivals, summer camps, plays, processions, socials, picnics, and programs. Through these religious activities, German-Russians—as Dakota Memories narrators attest—maintained traditions and adopted new customs.[74]

William Merkel Jr. shares his memories of church revival meetings, which are largely a Protestant activity. According to William's description, only the older members of the congregation participated in revivals. He remembers his parents, William Sr. and Magdalena (Joachim) taking part. At their Evangelical church, the revival meetings lasted for a few days, usually during weekends. William's parents only attended these meetings in the evenings, as they farmed during the days. Nonetheless, the revival meetings were community events. William maintains that all members, or at least "locals would attend these meetings."[75]

Churches also sponsored community socials or picnics. Growing up in a Roman Catholic family near Scott, Saskatchewan, during the 1930s, Anne (Frehlich) Hoffart has fond memories of attending church-sponsored picnics. Her congregation held them almost every Sunday afternoon, she says. They held these picnics immediately following mass. Members of the congregation participated, bringing a dish to the picnic and feasting throughout the afternoon. It was a picnic potluck, she recalls. Anne remembers that a game of ball usually broke out after a while.[76]

Dakota Memories narrators have vivid memories of the holiday programs sponsored by churches, such as the Christmas program and the Easter Procession. During their oral histories, narrators express joy when recounting their happy memories of religious holidays. Although some found the religious ser-

[74] Interviews with Axtmann, Backfish, Polly Berger, Wilfred Boechler, Merkel Jr., James Engelhart, Anne Hoffart, and Schauer.

[75] Interview with Merkel Jr.

[76] Interview with Anne Hoffart.

mon or mass drawn-out and boring, narrators still express having an appreciation for or an enjoyment in these programs.[77] For instance, Elsie Johnson remembers going to church for their Christmas program in their "Sunday best," a term narrators use to describe their finest clothes, often a special dress or suit reserved for church and important occasions. Each year her church hosted a program, consisting of its youngest members reciting German verses to their parents and community. Elsie remembers feeling quite anxious before the program, but by the end, she felt accomplished. Of course, she says, it always helped that the congregation gave each child a goodie bag that contained hard candy, ribbon candy, and an apple or an orange for participating in the program.[78]

While Elsie's memories center on Christmas, other narrators recall the significance of Easter celebrations. Anne (Frehlich) Hoffart, a native of Scott, Saskatchewan, has fond memories of her church celebrating Easter—Holy Thursday, Good Friday, and Easter Sunday. She remembers her church having grand processions in which four people carried "the statue of Christ risen" through the church and placed it on the altar. In the meantime, the congregation followed, taking their places in the pews, all the while singing beautiful religious hymns in German. These hymns, such as *Christ ist erstanden* ("Christ is Risen") and *Großer Gott, wir loben dich* ("Great God, We Praise You"), filled the church with a sense of reverence and celebration, reinforcing the deep spiritual meaning of the Easter season. The familiar melodies and solemn verses, passed down through generations, connected worshippers not only to their faith but also to their German-Russian heritage, making Easter a powerful and emotional experience.[79]

These German-Russian memory-makers insist religion was one of the most prevalent forces in their childhoods. While specific religious practices and beliefs vary from denomination to denomination, the role religion played in childhood was essentially the same for German-Russian Protestants and Catholics. Throughout the Dakota Memories oral histories, German-Russians from Russia of the second and third generations share memories of church preparations, services, rituals, and programs. These memories reveal the role of religion in shaping this ethnic group's identity. For instance, through their religious activities and celebrations, Germans from Russia were able to maintain their tra-

---

[77] Interviews with Axtmann, Backfish, Wilfred Boechler, James Engelhart, Anne Hoffart, Elsie Johnson, Merkel Jr., and Schauer.

[78] Interview with Elsie Johnson.

[79] Interview with Anne Hoffart.

ditions, while adopting new customs. By diffusing their past with their present, these German-Russian descendants were providing the necessary flexibility for their ethnic resilience.

Although much of rural childhood historiography overlooks the role of religion for children of ethnic groups, it is an essential category of analysis.[80] Scholars of ethnicity, like Pederson, recognize the role of religion in shaping ethnic identity.[81] German-Russian scholars, such as Kloberdanz and Arends, document the role of religion in the maintenance of German-Russian ethnicity.[82] Nonetheless, much of this literature insists that churches fostered either resistance or acceptance of assimilation.

By examining the past with nearly two hundred German-Russians of the second and third generations, Dakota Memories oral historians have revealed that churches, regardless of the denomination, fostered neither assimilation nor acculturation. Rather these oral histories show Germans from Russia using their religious institutions to blend their ancestral heritage with their new national identity. Churches provided a central place not only for worship but also for socialization, as congregants gathered before and after services to connect with neighbors, share news, and reinforce communal ties. Holiday programs, wedding receptions, and funeral gatherings often featured food traditions that further strengthened ethnic identity—whether it was the sharing of homemade kuchen at Easter or a community meal after a confirmation ceremony. Churches were not agents of change, however; rather the people themselves—particularly the younger generations—were responsible for shaping their evolving identity. Through this conservative institution, German-Russians absorbed bits and pieces of their new national culture while maintaining the familiar rhythms of faith, fellowship, and tradition that defined their community.

Born and reared near the small town of Venturia, North Dakota, in the 1940s, James Engelhart witnessed his church congregation at St. John's Catholic Church reshape his community's ethnic identity. "Religion was important during my childhood," he states. James remembers that most Sundays German-Russian farm families packed into St. John's for worship. "We just went to

---

[80] To review the lack of religion in the current rural childhood historiography, see West, *Growing up with the Country;* Riney-Kehrberg, *Childhood on the Farm;* Holt, *Children of the Western Plains.*

[81] For more information, see Pederson, *Between Memory and Reality.*

[82] More information can be found using Kloberdanz, "Volksdeutche: The Eastern European Germans"; Arends, *The Central Dakota Germans.*

church when our parents went." It was routine. He remembers partaking in a number of traditional activities—catechisms, confirmations, weddings, and funerals. Sermons and hymns, in James's church, were in German. Near the altar, James also remembers, stood two flags - an American flag and the church flag. These flags side-by-side, in a very traditional German-Russian institution, symbolize this community's willingness to blend American traditions with their ethnic traditions. This elasticity and determination resulted in the continuation of their ethnic identity, and, as a result, the creation of a revisited German-Russian story.[83]

---

[83] Interview with James Engelhart,

CHAPTER 8

Hope for Tomorrow: Holidays and Celebrations

"I don't remember Santa Claus," asserts Clara (Ell) Ferner. Born near Kronau, Saskatchewan, Clara only remembers celebrating Christmas once during her childhood, the year her parents took their five children on a long, tedious ride to visit Grandma and Grandpa Seiferling near Kerrobert, Saskatchewan, a small rural village 290 miles away. Christmas was a grand affair, a carefully orchestrated event filled with tradition and theatrics, as her maternal grandparents, Andrew and Katherine (Weimer), made the holiday unforgettable. The centerpiece of their celebration was a game they staged for the children, one that left a lasting impression. A frightening woman, the *Belznickel,* appeared, scolding the children for their misdeeds, her presence looming over them. The game was not without reward, however; before she left, the Belznickel recognized their good behavior and handed out small gifts. "To a small child, it made a big impression," Clara grins, the memory still vivid after all these years.[1]

This cherished memory of the Belznickel is just one memory Clara shares for the Dakota Memories project. Like other participants, she reflects on both hardship and celebration, recalling the harsh, unpredictable environmental—tornadoes, heat, wind chills, blizzards—alongside the unrelenting demands of farm life.[2] Yet, as Emmanuel Bertsch of Saskatoon, Saskatchewan, affirms,

---

[1] Interview with Clara (Ell) Ferner.

[2] Interviews with Nicholas F. Axtmann, Alice Ruth (Miller) Buck, Florence (Mitzel) Engelhart, Clara Ferner, Marvin Fiechtner, Stanley Helbling, Louis Helfrich, Ruth Rosa (Korus) Johnson, Eileen M. (Burlack) Ketterling, Rupert Kleingartner, Ida (Zimmerman) Koepplin, Viola (Eslinger) Konrad, Peter A. Kraft, Balzer Kurtz, Elizabeth Barbara (Rissling) Lang, Joseph Michael Meier, Irene (Bender) Schauer, and Arthur E. Schmidt.

the holidays and celebrations brought "hope for tomorrow."[3] These preserved childhood recollections of sacred and secular traditions offer a deeper look into the complexities of the German-Russian experience.

The Dakota Memories narrations show, for example, that slowly these immigrants increasingly borrowed and adapted American or Canadian customs. Historian Jon Gjerde, author of *The Minds of the West*, states that as immigrants "encountered institutional contacts, as they faced cultural intolerance in English publications, as they accepted material innovations at home, ethnic Americans—even those nestled deep in secluded communities—did indeed begin to change their behavior."[4] Determined to maintain their ethnic identity, Dakota Memories narrators reveal that while their ancestors conscientiously tried to maintain their ethnic customs, they also consciously accepted new ones.

In essence, the Dakota Memories narrations reveal "cultural diffusion."[5] As defined by anthropologist and folklorist Timothy J. Kloberdanz—who is also of German-Russia descent—cultural diffusion is "when one culture borrows certain elements from another."[6] He also refers to it as "cultural borrowing."[7] Kloberdanz points out that the adoption of American or Canadian practices is not representative of acculturation, but rather cultural diffusion. He says acculturation insinuates that mainstream society forced changes on an ethnic group, while cultural diffusion recognizes the free will and agency of a group. German-Russians appropriated the new customs, and thus it was cultural diffusion, not acculturation, taking place with the second and third generation.[8]

Maintaining the sacred holiday traditions was not just an act of faith but a means of preserving ethnic heritage and identity. As previously discussed in the chapter on religion, churches—both Protestant and Catholic—played a central role in sustaining cultural traditions, reinforcing community bonds, and providing continuity across generations. Jane Marie Pederson notes in *Between Memory and Identity*, that "besides the family, the institution which carried much of the responsibility for cultural transmission on the frontier and in rural areas was the church."[9] Beyond worship, these churches served as gathering

---

[3] Interview with Emmanuel Bertsch.

[4] Gjerde, *The Mind of the West*, 240; Clark, "Hope for Tomorrow."

[5] Kloberdanz, "The Traditions W Share," 30.

[6] Kloberdanz, "The Traditions We Share."

[7] Kloberdanz, "The Traditions We Share."

[8] Kloberdanz, "The Traditions We Share."

[9] Pederson, *Between Memory and Reality*, 116.

places hosting socials, picnics, prayer groups, and fundraisers that offered a welcome respite from daily hardships. Their most elaborate efforts, however, were reserved for holiday celebrations, where rituals, music, and fellowship brought the sacred to life, leaving lasting impressions on children and reinforcing the connection between faith and ethnic identity.[10]

German-Russians largely celebrated three sacred holidays: name days, Christmas, and Easter. Name days, tied to the feast day of a saint sharing one's name, often held more significance than birthdays and remained a largely sacred tradition. Christmas and Easter, while deeply rooted in religious observance, became increasingly secularized in America and Canada. These sacred traditions helped German-Russian immigrants maintain some of their cultural heritage, their religious beliefs, holidays, and foodways. "A sense of peoplehood," as ethnic scholar Pederson notes, "is deeply tied to a group's religion."[11] Thus, it is no surprise that the sacred holiday traditions of German-Russians are tied to their ethnic identity. Religion, including religious holidays, was a core element of their cultural preservation.

During their childhoods, Dakota Memories narrators did not realize the significance or importance of their ethnic traditions. They were just elated about celebrating a holiday. The purpose of the celebration was of little concern. As the narrations reveal, children looked forward to the sacred holidays because they could exchange a day of field labor for a day of socializing. These narrators reminisce more about eating, talking, and playing than about praying, honoring, and learning. For children a holiday was a holiday.[12]

For the purpose of this analysis, any holiday not celebrated as a religious celebration (or sacred holiday) by German-Russians is categorized as a secular holiday—a holiday detached from religious practices and beliefs. In America and Canada, German-Russians celebrated many holidays as secular: New Year's Day, Memorial Day, Mother's Day, Victoria's Day, Father's Day, Canada Day (Sports Day), Independence Day, Halloween, Thanksgiving, and birthdays. These secular holidays exposed ethnic groups, like German-Russians to a civic culture. "Rarely," as ethnic historian Jon Gjerde acknowledges, "were eth-

---

[10] Riney-Kehrberg, *Childhood on the Farm*, 148.

[11] Pederson, *Between Memory and Reality*, 116.

[12] Interviews with Alice Buck, Edwin "Emil" Dewald, Esther (Haerter) Drefs, Florence (Mitzel) Engelhart, Henrietta (Gall) Fiechtner, Calvin K. Fercho, ElRoy Fischer, Ruth (Dockter) Fischer, Martha M. (Brosy) Heyne, Anne (Frehlich) Hoffart, Lyla (Pressler) Johnson, Geraldine (Dukart) Keen, and Ida (Zimmerman) Koepplin.

nic communities so isolated and so consciously separate that they could avoid the influences of a quickening cultural evolution."[13] For instance, German-Russians lived in homogenous communities, yet often encountered outside influences through social institutions and individuals.

German-Russians on the northern plains adopted some American or Canadian secular holidays, blending them with their long-standing traditions while disregarding others. Independence Day—whether Fourth of July or First of July—was among the most celebrated, while holidays like Mother's Day, Father's Day, and Memorial Day were rarely mentioned. Their oral histories emphasize the holidays that brought the biggest celebrations and the most socializing, shaping their idea of a true holiday.[14] Regardless of whether a holiday was sacred or secular, children simply looked forward to the opportunity for fun. Independence Day, in particular, stood out as a favorite, offering a break from daily life and a chance for joy, celebration, and camaraderie. Through participation in these secular traditions, Germans from Russia increasingly saw themselves as American or Canadian Germans from Russia, reflecting their evolving identity in a new homeland.[15]

The blending of the ethnic traditions with civic customs might explain some of this German-Russian indifference to sacred and secular holiday traditions. While ethnic Americans, like Germans from Russia, willingly adopted some secular holidays, they completely ignored others or adopted bits and pieces of them. For instance, German-Russian immigrants slowly embraced the characters of Santa Claus and the Easter Bunny for their Christmas and Easter celebrations, respectively. Dakota Memories narrators express little concern regarding this transition. Yet, it caused confusion for German-Russian children. Rather than expressing concern, narrators simply stress the pleasure, joy, and happiness they found in these new holiday practices.[16] Since their parents and grandparents chose to accept and adopt these new customs and traditions, the second and third generations saw no reason to resist. Thus, the German-Russian identity was truly becoming a "rich mosaic" in the words of German-Russian scholar, Timothy J. Kloberdanz.[17]

---

[13] Gjerde, *The Minds of the West*, 230.

[14] Interviews with Backfish, Anna (Sturn) Doll, Emil Frank, and Robert R. Gab.

[15] Interviews with Magdalena Miller, Vernon Presler, Elda (Schultz) Rasch, Eugenia (Vetter) Schumacher, and Gorden R. Werth.

[16] Interviews with Emmanuel Bertsch, James Engelhart, Calvin K. Fercho, Clara (Ell) Ferner, Philip Lang, Gorden R. Werth, and Rueben Richard Wolf.

[17] Kloberdanz, "The Traditions We Share," 19.

The sacred holidays were the most important in the preservation of German-Russian ethnic identity, including customs and heritage. According to Dakota Memories narrators, name days were the only sacred holidays to remain uninfluenced by American or Canadian cultures. Name days were a Catholic celebration. Since name days were in honor of a saint, Protestant Germans from Russia did not celebrate them. Protestants were more likely to celebrate birthdays than name days. That said, not all Catholic narrators remember partaking in the name days festivities. "Customs vary not only from culture to culture," as anthropologist Timothy J. Kloberdanz states, "but also from family to family and individual to individual."[18] Nonetheless, those who did celebrate name days remember it as a celebration for people named after saints.[19]

Emmanuel Bertsch, who grew up in Scott, Saskatchewan, during the 1930s, insists this was an era when "we all wished we were saints." He points out that by celebrating name days the older generation was honoring tradition, as well as the saints. Emmanuel notes it was a time for people to gather and honor saints and those who shared the saints' names. During his oral history, he makes sure to point out there was "no Saint Emmanuel." He says it was a problem, but he "made up for it at Christmastime."[20]

The name-days festivities consisted of company, food, alcohol, and entertainment. It was largely a time of visiting and socializing. Depending on time and location, name days celebrations included immediate and extended family members, as well as close friends and neighbors. Red eye—a homemade or low-quality whiskey—and burnt whiskey, often made from overcooked or poorly distilled batches, were common at this celebration. The host family always provided plenty of delicious food, including some traditional German-Russian dishes. Narrators reminisce about feasting on large meals, usually dough-based, with lots of homemade sweets, such as *kuchen*. And, to top it off, there was usually entertainment. Dakota Memories narrators reminisce about live German music and card games. Name days were a time to celebrate.[21]

Wilfred Jacob Boechler, a native of Allan, Saskatchewan, has fond memories of this sacred holiday, especially the tradition of red eye liquor. He recalls

---

[18] Kloberdanz, "The Traditions We Share," 20.

[19] Interviews with Bertsch, Leo James Degenstein, Florence (Mitzel) Engelhart, Anne (Frehlich) Hoffart, Jacob Johner, Lawrence Johner, Geraldine (Dukart) Keen, and Fran Schwan.

[20] Interview with Emmanuel Bertsch.

[21] Interviews with Degenstein, Florence Engelhart, Anne Hoffart, Jacob Johner, Lawrence Johner, Keen, and Fran Schwan.

that even children were given a small shot in honor of the saints.[22] Leo James Degenstein, a Rugby, North Dakota, native, concurs with Wilfred, though he remembers honoring the saints with a shot of rye—a high-quality whiskey—rather than red eye.[23] Florence (Mitzel) Engelhart also cherishes memories of name days, recalling the red eye, food, and entertainment. Growing up near Strasburg, North Dakota, she remembers how the women and girls always prepared large meals for the celebrations. In the Mitzel house, meals were made in the basement, and carried upstairs, where guests feasted, listened to music, and played cards. Florence recalls a highlight was the homemade candies her mother often made for name days: divinity and fudge.[24]

While some narrators, like Wilfred, Leo, and Florence, have vivid memories of name days, others remember the custom dwindling away. There appears to be a generational divide or gap in regard to the celebration of name days. Per narrators this was largely a holiday for the older generations (grandparents and parents). German-Russian immigrants were no longer naming their children after saints; thus, a number of the second- and third-generation German-Russians do not recall name days as a part of their childhood. Some narrators remember the celebration taking place at night, while the children were in bed. Others remember being sent outside or to another room to play. The narrators also suggest that men may have practiced this tradition longer than women practiced it.[25]

Born in 1928, Leo James Degenstein remembers being sent to bed as soon as the name days festivities began, or at least his parents thought they were sending him to bed. According to Leo, the Degenstein children (boys and girls, young and old) always gathered in one of the upstairs bedrooms and spent the night huddled around a heat register. They were not cold. Rather, this heat register served as their conduit to the celebrations. Leo enjoys sharing this memory of eavesdropping. He has fond memories of listening to German music, adult conversations, and contagious laughter for hours upon hours.[26]

Although deeply committed to the preservation of their ethnic heritage, including their religious holidays, Germans from Russia encountered new cus-

---

[22] Interview with Wilfred Boechler.
[23] Interview with Degenstein.
[24] Interview with Engelhart.
[25] Interviews with Wilfred Boechler, Florence Engelhart, Degenstein, Jacob Johner, Lawrence Johner, and Keen.
[26] Interview with Degenstein.

toms in America and Canada that gradually shaped their identity. Exposure to different traditions led to adaptations, as this ethnic group incorporated elements of American and Canadian customs into their own celebrations. This blend of sacred and secular traditions is particularly evident in their observance of Valentine's Day, Easter, and Christmas. Dakota Memories narrations reflect both a steadfast commitment to sacred traditions and a willingness to embrace secular influences, illustrating the evolving and complex identity of the second- and third-generation German-Russians.[27]

Dakota Memories narrators assert that celebration of name days was untouched by their new surroundings; however, their memories of celebrating Valentine's Day are an exception to that assertion. Valentine's Day was one of the many name days. It was supposed to be a day to honor Saint Valentine. For Dakota Memories narrators, Valentine's Day was largely celebrated as a secular holiday. Only a few narrators even reference this day as a sacred holiday. In fact, it is more likely to hear a narrator reminiscing about celebrating Valentine's Day as a secular school holiday than a sacred church holiday.[28]

Valentine's Day was most commonly celebrated in the one-room country schools of the Dakota Memories narrators. This day was devoted to love expressed through handmade cards, candies, and small festivities. Some narrators have no recollections of celebrating, while others fondly remember crafting homemade Valentine cards to exchange with classmates. Store-bought cards were a rarity, with only a few narrators recalling ever using them. At school, children decorated Valentine boxes with construction paper, turning them into makeshift mailboxes for collecting cards and treats. Even teachers joined in the festivities, often bringing their own Valentines—cards, candies, and cookies—to share with students, making the day feel extra special.[29]

Henrietta (Gall) Fiechtner of Fargo, North Dakota, recalls celebrating Valentine's Day in a one-room country school near Wishek, North Dakota, where she made Valentine mailboxes out of old shoeboxes or something similar. In fact, she exclaims, the teacher even had prizes for the best mailboxes. After the teacher judged the mailboxes, Henrietta remembers using them to exchange

[27] Interviews with Bertsch, Wilfred Boechler, Alice Buck, Dewald, Esther (Haerter) Drefs, Florence Engelhart, Fercho, Henrietta (Gall) Fiechtner, ElRoy Fischer, Martha Heyne, Jacob Johner, Lyla (Pressler) Johnson, Geraldine Keen, Leona (Miller) Klundt, Ida Koepplin, and Kocher.

[28] Interviews with Henrietta (Gall) Fiechtner, Geraldine (Dukart) Keen, and Koepplin.

[29] Interviews with Henrietta Fiechtner, Keen, and Koepplin.

Image 8.1. Marlene Ann (Kraft) Smith's class celebrating Valentine's Day, *circa* 1946–1947. This photograph highlights how holiday festivities brought joy, strengthened friendships, and connected rural German-Russian children to wider American traditions. Back row (left to right): Eyvonne (Kraft) Guillen, Genevieve (Kraft) Petry, and Opal Sather. Middle row (left to right): Howard Thompson, Marlene Ann (Kraft) Smith, Richard Kraft, and Marla Sather. Front row (left to right): Bonnie Sather, Irma Thompson, and Boyd Sather. *Photograph from the Marlene Ann (Kraft) Smith Collection, GRHC.*

Valentines. During her oral history session, Henrietta also makes sure to acknowledge the generosity of her teacher. Not only did her teacher provide a prize for the best mailbox, but she also brought candy. According to Henrietta, her teacher always had little heart-shaped candies for the students to enjoy.[30]

Dakota Memories narrators spend a good deal of time sharing their memories of Easter. The second- and third-generation German-Russians reveal that Easter was a religious holiday with a secular twist. That is, narrators reminisce about celebrating Easter at church, home, and school. The church celebration varied largely from religion to religion, as well as church to church. Nonetheless, the Easter celebration maintained its religious focus. Since this holiday was

[30] Interview with Henrietta Fiechtner.

in honor of Jesus Christ's resurrection, it had a special place in the sacred traditions of Germans from Russia. Exposure to American and Canadian practices, however, changed the celebration of Easter in numerous German-Russian homes, as well as schools. Slowly, as well as increasingly, Germans from Russia began to adopt more secular customs, like the Easter Bunny and colored eggs.[31]

For German-Russians, the traditional Easter celebration began at church. Protestants largely celebrated on Easter Sunday, while Catholics largely celebrated over the course of three or four days. Actually, some Catholic narrators insist the holiday started with Lent, forty days before Easter Sunday. For most narrators, however, it consisted of Holy Thursday, Good Friday, and Easter Sunday. Some Catholics recall there being an Easter vigil on Saturday, as well. Regardless of whether occasion lasted one morning or forty days, German-Russian descendants associate the Easter holiday with church services or festivities.[32]

According to Wilfred Jacob Boechler, a devout Catholic, Easter was a religious holiday that lasted multiple days. He calls it the Easter Triduum. The holiday comprised Holy Thursday, Good Friday, and Holy Saturday. Sitting in a pew in the Catholic Church of Allan, Saskatchewan, Wilfred explains the Triduum. He remembers that Holy Thursday was in honor of Jesus's last supper with his disciples. To celebrate Holy Thursday, the Boechlers attended a two-hour mass. The priest, as Wilfred recalls, provided children with a special lesson on morals during the Thursday worship. Good Friday, as Wilfred explains, commemorated the death and crucifixion of Jesus. Again, Wilfred remembers attending church on this day. On Holy Saturday, the Boechlers went to church yet again. This time, as Wilfred points out, they partook in an Easter vigil in anticipation of Jesus's resurrection. And, last but not least, Wilfred remembers attending church on Easter Sunday. Wilfred acknowledges that for young children this holiday seemed important, but never-ending.[33]

Catholic narrators also remember taking part in an Easter Procession. Depending on one's locality, the procession took place in town or near the country church or school. Typically, the procession involved the congregation meeting at one central location, such as a parish schoolhouse, and marching together

[31] Interviews with Bertsch, Wilfred Boechler, Alice Buck, Dewald, Drefs, Florence Engelhart, Fercho, ElRoy Fischer, Ruth Fischer, Martha Heyne, Anne Hoffart, Jacob Johner, Lawrence Johner, Adam A. Kaseman, Keen, Lorraine Kleingartner, Klundt, Peter Kocher, and Ida Koepplin.

[32] Interviews with Bertsch, Wilfred Boechler, Fercho, iAnne Hoffart, Jacob Johner, Keen, and Klundt.

[33] Interview with Wilfred Boechler.

toward the church. During the march, as some recall, the congregation sang hymns and participated in prayers. During his oral history session, Emmanuel Bertsch of Saskatoon, Saskatchewan, reminisces about his sore knees. He complains, "I was just a kid!"[34] Some narrators also recall a member of the congregation carrying a statue of Jesus Christ. For countless German-Russian immigrants, this religious procession was another means of honoring their religious convictions and traditions.[35]

In America and Canada, however, this religious celebration was accompanied by secular traditions, such as the Easter Bunny, Easter baskets or nests, colored eggs, and Easter Egg Hunts. According to the folklore of the second- and third-generation German-Russians, an Easter Bunny would usually appear late at night, after all children were in bed and asleep, and leave behind gifts. Some remember the Easter Bunny leaving goodies and candies—chocolate bunnies to be specific—oranges, apples, and nuts, in makeshift baskets. Others remember making nests for the Easter Bunny, and leaving them at the foot of the bed or outside near the barn. Only a few narrators remember having or receiving store-bought baskets or materials like cellophane grass.

Alice Buck of Streeter remembers celebrating Easter as a sacred-secular holiday. Growing up a Lutheran girl, Alice recalls going to church on Easter Sunday. Her favorite memory of this holiday, however, has nothing to do with religion. Alice glows with excitement recalling one of her favorite childhood memories. It was the Easter her family spent at Grandmother Miller's place. That Easter, Alice's mother, Christina (Kammerer), had the children set out their baskets for a bunny. Seemingly surprised by this notion, Alice explains that this bunny was supposed to leave goodies for children in these baskets. When Leona, Alice, and Norman, the three Miller children, awoke, they found their baskets empty. Their mother insisted, as the story goes, that she saw a "nest" in the barn. According to Alice, "She had gotten a bunch of rabbit turds, and . . . put them into a little nest of hay, and she said 'there, that's what the Easter Bunny brought you.'" Alice remembers her trying to convince them that they were chocolate covered peanuts. "We knew better," Alice exclaims. When they returned to the house, Christina gave each of her children a chocolate covered Easter Bunny.[36]

---

[34] Interview with Bertsch.

[35] Interviews with Bertsch and Anne Hoffart.

[36] Interview with Alice Buck.

Image 8.2. Marilyn Jeanette (Sauer) Bauman's class with their Easter baskets, ***circa*** **1940s. This joyful scene illustrates how school celebrations reinforced community bonds, religious identity, and cultural pride among German-Russian children. Marilyn is the second from the right.** *Photograph from the Clarence and Marilyn Jeanette (Sauer) Bauman Collection, GRHC.*

Coloring eggs was another highlight for German-Russian children. This secular activity mostly took place at home, although some narrators reminisce about coloring eggs and making nests or baskets at school. This activity was usually, but not always, done by the children. Some narrators remember their mothers and older siblings always did the coloring late at night, allowing the children to be astonished by seeing colored eggs in their nests. Other narrators remember using dye, either homemade, like beet juice, or store-bought. However the colored eggs came to be, the so-called Easter bunny left them in Easter baskets or nests at night with the other goodies.[37]

As with many of the favorite German-Russian holidays, Easter largely involved visiting and socializing. Depending on the weather, a typical Easter celebration included the immediate family and extended family, as well as close neighbors and friends. The celebration usually centered on feasting and

---

[37] Interviews with Bertsch, Alice Buck, Drefs, Ruth Fischer, Anne Hoffart, Jacob Johner, Lawrence Johner, Lyla Johnson, and Ida Koepplin.

children's activities. Dakota Memories narrators reminisce about the big meal, usually consisting of dough-based dishes and sweets, but some some narrators insist Easter dinner was nothing out of the ordinary. For those who partook in the large feast, they usually took part in some games and activities as well, such as Where's the Easter Basket? and Easter Egg Hunts. Who Has the Strongest Egg was a traditional egg-tapping contest where participants would tap their hard-boiled eggs against each other to see whose egg remained uncracked the longest. Where's the Easter Basket? was a playful scavenger hunt, with children searching for hidden Easter baskets filled with treats. Easter Egg Hunts, a familiar tradition, involved children racing to find brightly colored eggs hidden around the home or yard, a favorite part of the celebration. These activities kept children entertained for hours. They, as well as the other secular activities, are also the reason Dakota Memories narrators think so highly of this holiday. The secular activities associated with Easter, which provided an opportunity for socializing, were especially fun for the children.[38]

Christmas was the holiday Dakota Memories narrators recall as both secular and sacred. Narrators recall a schedule that include midnight Mass, evening supper, the Belznickel, the *Krist Kindel* (the Christ Child), gifts, church and school programs, and decorations. For narrators, this holiday was by far the favorite. In addition, the collective memory of this holiday reveals two generations in the midst of cultural diffusion, as some children have memories of the Belznickel and Krist Kindel, and some have memories of Santa Claus. Moreover, a few reveal memories of all three.[39] This holiday crossed the sacred-secular realms. It was, as narrators attest, celebrated in churches, schools, and homes.[40]

German-Russian descendants grew up well aware that Christmas was a religious holiday. Congregations typically held the festivities during a two-day period, Christmas Eve and Christmas Day, although most churches—regardless of the denomination—began preparing for the festivities well before Christmas Eve. Dakota Memories narrators remember their churches being decorated from top to bottom. There were Christmas trees and nativity scenes. Countless children grew up amazed by the beauty of their church at Christmastime.

---

[38] Interviews with Drefs, Martha Heyne, Keen, Klundt, and Ida Koepplin.

[39] Clark, "Hope for Tomorrow."

[40] Interviews with Axtmann, Bertsch, Alice Buck, LeRoy Buck, Dewald, Stanley Graven Dobler, Drefs, Florence (Mitzel) Engelhart, Fercho, Ben Ferner, Clara Ferner, Henrietta Fiechtner, Marvin Fiechtner, ElRoy Fischer, Ruth Fischer, Leo Peter Herauf, and Herman.

Church services usually involved a midnight worship or Mass on Christmas Eve, and a morning service or Mass on Christmas Day.[41]

Churches, as the Dakota Memories narrations reveal, also held some type of holiday program during the week of Christmas. Some programs were in conjunction with the church services (like midnight Mass or service on Christmas Eve or morning Mass or service on Christmas Day); other churches held the program a week or so before Christmas. The Christmas programs varied from church to church. Nonetheless, it seems that recitations, skits, goodie bags, and Christmas hymns were common practices. It was truly a community event or celebration. Memories of these programs are etched in the minds of countless German-Russians on the northern plains.[42]

Given that schools on the northern plains implemented a holiday break in winter, schools typically celebrated Christmas a few weeks before the actual holiday. When thinking about Christmas celebrations in school, narrators recall the holiday break, decorations, and school programs. As children, narrators remember feeling conflicted about the holiday break. On the one hand, they had no school for a few weeks and Christmastime was fast approaching; on the other hand, no school usually meant more chores and responsibilities around the house and farmstead. Before school let out for the holiday recess, children participated in making and hanging Christmas decorations throughout the schoolhouse. Some narrators even remember their schoolhouses having Christmas trees, a rare sight to see on the northern plains at that time. These narrators also reminisce, and some at length, about holding a Christmas program or concert at school. Sometimes this event was simply a community celebration; sometimes it was a school fundraiser.[43]

Alma (Janke) Schott, from Gackle, North Dakota, remembers that the community participated in all school programs, including the Christmas program. There was never an empty seat, she says. Alma remembers her teacher always had the schoolhouse frugally decorated and a Christmas tree ornamented with

[41] Interviews with Axtmann, Bertsch, Alice Buck, Florence Engelhart, Fercho, Ben Ferner, Henrietta Fiechtner, ElRoy Fischer, Ruth Fischer, Martha Heyne, Anne Hoffart, Lawrence Johner, Lorraine Kaseman, Peter Kocher, and Joseph George Kohlman.

[42] Interviews with Alice Buck, Drefs, Marvin Fiechtner, ElRoy Fischer, Ruth Fischer, Martha Heyne, Anne Hoffart, Ruth Johnson, Adam Kaseman, Lorraine Kaseman, Rupert Kleingartner, Klundt, and Peter Kocher.

[43] Interviews with Rupert Kleingartner, Keen, Alice Buck, Klundt, Ida Koepplin, and Kohlman.

student-made paper chains.[44] Joseph George Kohlman, from Tramping Lake, Saskatchewan, also has fond memories of his school's Christmas program and concert. He remembers participating in the school play every year. One year, as he recalls, he was the letter "C" in a recitation program which required all the children to be responsible for a letter in the word Christmas. "C is for Christmas with lights so bright," Joseph utters, as he is swept up in the memory.[45]

The best part of the Christmas holiday, at least according to Dakota Memories narrators, was the celebration that took place at home. During their oral history sessions, narrators spend a good deal of time reminiscing about the secular activities that took place in their childhood homes. These narrators have fond memories of family gatherings, Christmas characters—the Belznickel, Krist Kindel, and Santa Claus—in addition to gifts, decorations, feasts, and activities. Without a doubt, the festivities that took place in the homes of narrators made this holiday their favorite.[46]

Families, immediate and extended, gathered to celebrate Christmas, weather permitting of course. A Christmas family scene typically involved socializing through pastime activities. Depending on the family, Christmas gatherings took place on either Christmas Eve or Christmas Day. Dakota Memories narrators reminisce about playing games, singing songs, and reading stories. For instance, while the adults played card games like Whist—a trick-taking game that required strategy and teamwork—the children played race games, like Tag, where one player chased others to "tag" them, and Hide 'n' Seek, where one child closed their eyes and counted while others hid, trying not to be found.

Singing Christmas carols was also a favorite pastime. Narrators recall singing "Up on the House Top" and "Silent Night." Some, however, chose to listen to Christmas music via the radio instead of singing. Reading the Christmas story from the Bible was also a favorite pastime, as well. Usually an adult member, like a grandparent or parent, would read aloud, while the rest of the family, adults and children alike, gathered around to listen. Christmas was a time for family members to reconnect and enjoy one another.[47]

---

[44] Interviews with Alma (Janke) Schott; Clark, "The German-Russian Christmas: Oral Histories from the Northern Plains," *Heritage Review* (2007): 8.

[45] Interview with Kohlman.

[46] Interviews with Alice Buck, LeRoy Buck, Dewald, Stanley Graven Dobler, Drefs, Florence Engelhart, Fercho, Clara Ferner, Marvin Fiechtner, Lyla (Pressler) Johnson, Ruth Johnson, Keen, and Balzer Kurtz.

[47] Interviews with Alice Buck, LeRoy Buck, Dobler, Florence Engelhart, Henrietta Fiechtner, ElRoy Fischer, and Keen.

Image 8.3. Dakota Memories participant Lillian Elsa (Oster) Gebhardt's Christmas celebration with Santa Claus, date unknown. This moment captures the joy and warmth of holiday traditions that blended German-Russian customs with American influences, strengthening family and community bonds. *Photograph from the Ervin and Lillian Elsa (Oster) Gebhardt Collection, GRHC.*

At these family gatherings, there were impressive feasts. A few narrators do not remember anything special for Christmas dinner, but others insist the meals were atypical of a regular meal. Some families held the Christmas dinner on Christmas Eve after church; other families held it on Christmas Day after church. Countless narrators remember feasting on a variety of meats: sausage, smoked ham, turkey, and goose. They also reminisce about the ethnic dough dishes: strudels, *kase knephla*—German-Russian dish of dumplings filled with cheese and often served with butter or cream—and *fleischkuechle*, a savory, deep-fried pastry stuffed with seasoned ground beef. And, of course, narrators remember all the sweets: candies, cut-out cookies, kuchen, home-made ice cream, fruits, and pies. Feasting was simply part of the Christmas celebration.[48]

The Kohlmans, as Joseph George Kohlman recalls, always enjoyed their Christmas feast after midnight Mass. He fondly remembers the mouthwater-

[48] Interviews with Bertsch, Alice Buck, Dobler, Drefs, Marvin Fiechtner, ElRoy Fischer, Ruth Fischer, Keen, and Kohlman.

ing Christmas meals of his childhood, which typically featured a stuffed goose, bread, milk, eggs, apples, and raisins. But the highlight, he proclaims, was the crispy, burnt skin of the Christmas goose—a delicacy that sparked friendly competition among family members. Every year, they eagerly vied for a piece, especially the coveted crunchy skin from the goose's neck and breast. Even during the interview, Joseph admits he gets hungry just thinking about it.[49]

The Christmas celebration commonly included a traditional character or characters. For the second and third generation, these characters included the Belznickel, the Krist Kindel, Santa Claus, or any combination of those three. The Belznickel was a mean, scary character. He was traditionally a male member of the family dressed in a fur coat and hat, carrying chains and a whip. He was responsible for punishing the naughty children. The Krist Kindel was a beautiful, charming character. She was traditionally a female family member dressed in white with a veil. She was responsible for rewarding children for their good deeds. Santa Claus was a jolly male character who mysteriously visited homes late at night, after all the children were in bed and fast asleep, and left behind presents.[50]

Growing up near Devils Lake, North Dakota, Balzer Kurtz remembers all three: the Belznickel, the Krist Kindel, and Santa Claus. According to Balzer, the festivities began two weeks before Christmas. As he remembers it, first there was the Belznickel: "He was a mean dude," exclaims Balzer. The Belznickel came to identify which children had been naughty and which had been nice. Balzer recalls that the Belznickel always reeked of booze, carried a willow whip, and dragged a chain. Then, a few days later, the Krist Kindel appeared. Balzer recalls the Krist Kindel always being a woman—a woman dressed in colorful formal attire, with a hat and veil. She was the nice one. She gave the children candy. Then, on Christmas Day, Santa Claus came to the house. As Balzer recalls, on Christmas morning someone always came to his house dressed up as Santa Claus with a small gift for every child. [51] For young children, these colorful characters were the highlight of Christmas.[52]

---

[49] Interview with Kohlman.

[50] Sherman, *Plains Folk*, 152; "An Old-Fashioned Christmas," *Prairies* (1983–1984); Warren Overlie, "A Russian Christmas," *Prairies* (1984-1985); "There were no Christmas Tress for Germans from Russia," *Prairies* (Winter 1978).

[51] Interview with Kurtz.

[52] Clark, "The German-Russian Christmas," 9.

Besides the Christmas characters, German-Russians also remember decorations and gifts. Some families had Christmas trees, while others had none.[53] Born in Gackle, North Dakota, Emil E. Schaffer remembers "always [having] a Christmas tree." He also acknowledges that decorations for the tree were sparse, usually limited to popcorn strings and homemade ornaments.[54] Since second- and third-generation German-Russians grew up in rural areas without electricity, Christmas lights were rare. Some families, like Elizabeth Barbara (Rissling) Lang's family, simply did without; other families, like Balzer Kurtz's family, used candles.[55] When looking back on their childhood, German-Russians realize not only did they have few decorations, but they also received few gifts. As Balzer states, "Christmas was a handful of peanuts, a little Christmas candy, and an orange or an apple, if [our parents] could afford it. That was it."[56] Christina Rose Marie (Bast) Krismer recalls getting a dollar bill, a chocolate bar, and a knife, fork, or spoon.[57] As Emil E. Schaffer says, "we got what we needed."[58]

The second- and third-generation German-Russians also have vivid memories of partaking in a few distinctly secular or civic holiday traditions, such as Independence Day (or Canada Day) and birthdays. These holidays usually involved a community (or school) celebration, and were usually part of the "cultural borrowing" process.[59] Typically held by a town or school, these celebrations were distinctly American or Canadian. These celebrations spread to the isolated, rural areas, and became part of German-Russian life on the northern plains. They brought people together for common purposes, relief, and enjoyment.[60]

Mumming on New Year's Day, for instance, was a common celebration, although Dakota Memories narrators never use this term. Mumming, or visiting, was a tradition among neighbors. Some Dakota Memories narrators reminisce about celebrating on New Year's Eve; some remember celebrating on New Year's Day; others have no recollections of celebrating this holiday. Regardless, it seems that for Dakota Memories families the New Year's celebration

---

[53] Joseph Height, "Christbaum and Christkindel," *North Dakota Historical Society of Germans from Russia Heritage Review* 16 (1976): 4.
[54] Interview with Emil E. Schaffer.
[55] Interviews with Kurtz and Elizabeth Barbara (Rissling) Lang.
[56] Interview with Kurtz.
[57] Interview with Krismer.
[58] Interview with Schaffer; Clark, "The German-Russian Christmas," 9.
[59] Kloberdanz, "'The Traditions We Share," 30.
[60] Clark, "Hope for Tomorrow."

Image 8.4. Dakota Memories participant Herbert Herman wearing his little "soldier" suit, sitting next to his aunt Bertha (Roloff) Herman, *circa* 1919. Herbert remembers wearing this suit on New Year's Day when he visited neighbors to wish them a Happy New Year. This tradition strengthened social bonds and reflected the pride families placed in teaching community values. *Photograph from the Herbert Herman Collection, GRHC.*

was a holiday for children, children of all ages. The tradition of mumming was largely reserved for younger children. However, older children found fun in participating in this tradition, as well. Narrators reminisce about firecrackers, guns, pranks, and alcohol. Sometimes the younger children found this holiday overwhelming.[61]

Born in 1915, Herbert Herman recalls celebrating New Year's Day as a young lad in Gackle, North Dakota. During his oral history session, Herbert shares his first memory of celebrating New Year's Day in Gackle. To help him adjust to town life, Herbert's mother, Otillie (Roloff), encouraged Herbert to partake in the town activities, such as mumming. Herbert's mother dressed

[61] Interviews with Drefs, Herman, Peter Kocher, and Ida Koepplin.

him in a little "soldier" outfit, gave him a cork-gun, and sent him visiting. He remembers his mother teaching him a short German piece to recite at each home, which he can still recite with ease and clarity.[62] Other narrators, such as Ida (Zimmerman) Koepplin of Lehr, North Dakota, and Peter Kocher of Hague, North Dakota, recall people using real shotguns and firecrackers to celebrate the New Year. According to them this celebration involved visiting, drinking, and lots of noise.[63] Ida insists, "It was a scary night."[64]

During their oral histories, Dakota Memories narrators rarely mention the next secular holidays on the calendar: Mother's Day, Memorial Day, Victoria Day, and Father's Day. Given that these holidays did not involve a large town or school celebrations on the northern plains, they are largely forgettable. Still, a few narrators, such as Lyla (Pressler) Johnson and Marlene Ann (Kraft) Smith, share memories of these traditions, even though they are brief and vague.[65] Narrators seem to stumble onto these memories. For instance, Lyla's memories of Memorial Day had more to do with her being a Girl Scout than the actual celebration. She has fond memories of dressing up in her Girl Scout uniform and marching to the cemetery with her fellow scouts.[66] Moreover, Marlene only shares her memories of Mother's Day during a photo album tour. Once Marlene saw a photograph of a Mother's Day celebration during her childhood, she began to share her memories of this holiday.[67] Thus, it appears that these holidays, Memorial Day and Mother's Day, are not as memorable as other secular holidays.

Independence Day (Fourth of July or First of July) is quite a different story. Celebrating this holiday was the highlight of children's lives. Dakota Memories narrators spend just about as much time reminiscing about this secular holiday as they do reminiscing about the sacred holiday, Christmas. Independence Day was truly a community celebration. There were festivities in all major towns, as well as most small, rural towns. Narrators recall picnics, parades, ball games, races and competitions, dances, fireworks, and carnivals. It was,

[62] Interview with Herman.

[63] Interviews with Kocher and Ida Koepplin.

[64] Interview with Ida Koepplin.

[65] Interviews with Lyla (Pressler) Johnson and Marlene Ann (Kraft) Smith.

[66] Interview with Lyla Johnson.

[67] Interview with Smith.

for German-Russian families, the perfect break from the monotony of daily life and summer chores.[68]

Since children looked forward to this holiday, and it was during the busiest season for farmers, parents used this holiday to inspire production. According to the Dakota Memories participants, attending the Independence Day celebrations was a reward for good behavior. The fond memories of this holiday probably correlate with the idea that participation in the festivities was a reward. As Emil Dewald points out, "We couldn't go to Fourth of July until we had our chores done."[69] Many narrators recall their parents having a similar rule.[70]

Nonetheless, Dakota Memories narrators have fond memories of celebrating the birth of their nations. For the most part, narrators remember attending large community celebrations. Sometimes these town festivities lasted only a day or evening, while other times they lasted two to three days. The longer festivities usually took place in larger towns and in conjunction with another event, such as a rodeo or carnival. Regardless, there were typically town parades in honor of these days.[71] Father Leonard A. Eckroth has fond memories of this holiday. He recalls the Fourth of July festivities typically lasting three days, as they included a parade and rodeo in town. Wearing his priestly garb, and a big grin, Father Eckroth makes a point to share his memories of his father buying and setting off firecrackers with the children.[72]

Firecrackers were another childhood favorite from the Fourth of July or First of July celebrations. Countless Dakota Memories narrators, like Father Eckroth, have fond memories of playing with firecrackers.[73] A few, like James Engelhart, have memories of watching large firework shows.[74] Some narrators, like Calvin Fercho, remember the dangers of firecrackers, as well. Despite all the fun and games, these celebrations occasionally became tragic, as Calvin recalls. Growing up near Lehr, North Dakota, in the 1930s, Calvin remembers investing all his "spare money . . . in firecrackers." He and his friends found

[68] Interviews with Backfish, Doll, James Engelhart, Fercho, Emil Frank, Robert R. Gab, Jacob Helbling, Martha Heyne, Koepplin, Philip Lang, Magdalena Miller, Eva (Helfrich) Mutschelknaus, Vernon M. Presler, Elda (Schultz) Rasch, Orion Arlyn Rudolph, Eugenia (Vetter) Schumacher, Simon Schwan, Gorden R. Werth, and Rueben Richard Wolf.

[69] Interview with Edwin "Emil" Dewald.

[70] Interviews with Dewald, ElRoy Fischer, and Martha Heyne; Clark, "Hope for Tomorrow."

[71] Interviews with Alice Buck, Dewald, Fr. Leonard A. Eckroth, Florence Engelhart, interview; Herbert Herman, and Lorraine (Meidinger) Kaseman.

[72] Interview with Fr. Leonard A. Eckroth.

[73] Interview with Eckroth.

[74] Interview with James Engelhart.

firecrackers quite entertaining, especially when placed inside glass bottles. They had more fun watching the glass explode. Their fun continued, until the one time the explosion sent a shard of glass into his cousin's face, just above the eye. Calvin, who as an adult became an ophthalmologist, insists it was pure luck his cousin was not seriously injured.[75]

The town or community celebrations were particularly fun because of all the activities available for children. Dakota Memories narrators have memories of horse races, bicycle races, foot races (such as sack races and three-legged races), tug-of-war, and ball games.[76] At Salt Lake (an alkaline lake in south-central North Dakota), there was roller skating, air plane rides, dances, boat rides, horse shoe tournaments, horse races, foot races, parachuting pilots, and swim competitions.[77] Some narrators remember guest speakers or entertainers coming to town, too, like Gene Autry.[78] There were also big Fourth of July or First of July dances for people of all ages.[79]

Families, as Dakota Memories narrators attest, spent the entire day in town enjoying the festivities. A midday homemade picnic was common, typically consisting of sandwiches; many narrators specifically recall bologna sandwiches or bread and sausage with soda. Some remember their parents giving them spending money for a hot dog and some sweets. Beyond food, many narrators fondly recall spending their extra change on dresses, haircuts, ice cream, fountain sodas, candy-like jawbreakers, and crackerjacks.[80] Martha M. (Brosy) Heyne, from Wishek, North Dakota, vividly remembers one year when she "got fifty cents and that was a lot of money." Dissatisfied with her thick, heavy hair, she decided to visit a beauty shop, where the "beauty operator" charged her only a quarter. So, she got her hair cut. She recalls she "had a quarter to spend yet." That remaining twenty-five cents, she says, "was enough for whatever I wanted."[81]

A few Dakota Memories narrators remember participating in a farm celebration rather than attending a large civic celebration in town. The farm celebration was modest in comparison to the town or community celebrations.

---

[75] Interview with Fercho.
[76] Interview with James Engelhart.
[77] Interview with ElRoy Fischer.
[78] Interview with Simon Schwan.
[79] Interview with ElRoy Fischer.
[80] Interviews with Drefs, Eckroth, Fercho, Henrietta Fiechtner,Herman, Martha Heyne, Lyla Johnson, Adam A. Kaseman, Lorraine Kaseman, and Leona (Miller) Klundt.
[81] Interview with Martha Heyne.

Nonetheless, German-Russian farm children remember it involved visiting with family, neighbors, and close friends. The festivities included feasting on delicious homemade foods, like ice cream, soda, kuchen, potato salad, and chicken; setting off firecrackers; and enjoying barn dances.[82] Jacob A. Helbling of Bismarck, North Dakota, remembers having barn dances in his family's hay loft. "The Steckler boys [musicians]would come to play," Jacob recalls, "They had a huge crowd by two o'clock." They played for fifty to seventy people, Jacob insists, and "[t]he dance would sometimes go until four in the morning."[83]

Halloween and Thanksgiving were not quite as memorable as Fourth of July and Canada' First of July for Dakota Memories narrators. In fact, the narrator responses regarding these holidays are mixed. Some remember celebrating them; others insist these holidays were not part of their childhood. Those who celebrated Halloween remember it largely as a school celebration; those who celebrated Thanksgiving recall it was a family gathering. Regardless, a few German-Russian children found pleasure in these holidays.

The Halloween traditions involved costumes and pranks. Some Dakota Memories narrators remember going to school all dressed up in homemade costumes or funny outfits. A few remember costume competitions. Geraldine (Dukart) Keen of Saskatoon, Saskatchewan, recalls having Halloween parties at her country school. According to her, children always dressed up in costumes, usually made by their mothers, or "funky" outfits.

Halloween was also an occasion for childhood pranks and jokes.[84] As novelist Lois Phillips Hudson writes, "Nobody in . . . North Dakota had ever heard of trick-or-treating, that apologetic form of annoying neighbors, subsidizing candy-makers, ruining teeth, upsetting stomachs, and cheating red-blooded human spirits."[85] In her short story "The Buggy on the Roof," Hudson elaborates on the traditional Halloween celebration of the 1930s.[86] Likewise, Ida (Zimmerman) Koepplin recalls, Halloween was celebrated by playing pranks, like outhouse tipping, rather than trick or treating.[87]

---

[82] Interviews with Jacob Helbling and Rupert Kleingartner.

[83] Interview with Jacob Helbling.

[84] Interview with Geraldine (Dukart) Keen.

[85] Hudson, *Reapers of the Dust,* 39.

[86] Hudson, *Reapers of the Dust,* 39–44

[87] Interview with Koepplin.

Image 8.5. Alamo School children dressed up for a Halloween party, 1949. Costumes reflected popular themes of the time and were chosen in a spirit of fun and imagination, though they also illustrate cultural attitudes that were common or representative of the era. This photograph shows how rural German-Russian communities embraced American holiday traditions to create joyful, shared experiences. *Photograph from the Clarence and Marilyn Jeanette (Sauer) Bauman Collection, GRHC.*

Unlike Halloween, Thanksgiving was largely a family celebration. It involved, as the Dakota Memories narrations reveal, visiting, socializing, and feasting. Family and friends would travel great distances, weather permitting, to attend Thanksgiving dinners. This holiday was simply an excuse for people to gather and socialize. In Canada it was celebrated in the middle of harvest season; in the United States it was celebrated at the onset of winter. Regardless of the October or November timing, Thanksgiving was all about the food—the turkey, dressing, and cranberries—and the family. Only a few narrators, like Stanley Graven Dobler, label Thanksgiving as particularly memorable.[88]

Another secular holiday that German-Russians Russia willingly adopted was the celebration of birthdays, although not all German-Russians remember celebrating their date of birth. As previously mentioned, Catholic Germans from Russia were still practicing name days during this time. Thus, Catholic narrators have memories of celebrating name days instead of birthdays. Never-

[88] Interviews with Stanley Graven Dobler, Henrietta Fiechtner, Lyla Johnson, Keen, and Koepplin.

theless, there are a few Catholic narrators and a number of Protestant narrators who remember celebrating birthdays.[89]

The birthday celebrations, as the Dakota Memories narrations reveal, were rather small, intimate gatherings. Usually, they only consisted of the immediate family, parents and children. Families usually held them only for younger children below the age of ten. A birthday celebration involved a special dinner (usually the birthday boy's or girl's favorite meal), followed by homemade cake and ice cream. Some families gave gifts to the birthday boy or girl, but usually the gifts (or gift as the case may be) were small and practical. Rarely did children receive luxury items, such as toys, trinkets, or games. For instance, Fran Schwan, from Rugby, North Dakota, remembers getting a lamb for her fourteenth or fifteenth birthday. She named it Liberace.[90]

For children of German-Russian immigrants, holidays sacred and secular, provided an immense amount of enjoyment. As Dakota Memories narrators attest, they never really cared if the holiday was sacred or secular, traditional or modern. They simply enjoyed the celebration. The opportunity for socializing, visiting, feasting, and playing elated them. It was pure fun. These narrators were not concerned about such celebrations causing them to lose or change their ethnic identity. For children, a holiday was simply a holiday.

Nonetheless, these holiday traditions, the sacred, secular, and mixture of sacred-secular, reveal the efforts of German-Russian immigrants to maintain their ethnic identity, as well as their efforts to become American or Canadian. As Kloberdanz states in "The Traditions We Share," Germans from Russia chose to borrow certain elements (such as some secular holidays) from their new cultural environment.[91] They never abandoned their ethnic identity. They altered it. Thus, German-Russian descendants have a complex identity, a rich ethnic mosaic.

As adults looking back on their childhood, Dakota Memories narrators are complicating the traditional German-Russian story. They are actively shaping

---

[89] Interviews with Bertsch, Wilfred Boechler, Alice Buck, Dewald, Dobler, Drefs, Florence Engelhart, Fercho, ElRoy Fischer, Ruth (Dockter) Fischer, Anne (Frehlich) Hoffart, Jacob Johner, Lawrence Johner, Lyla Johnson, Keen, Rupert Kleingartner, and Klundt.

[90] Interview with Fran Schwan.

[91] Kloberdanz, "The Traditions We Share," 30.

it by participating in Dakota Memories. According to these narrators, the holidays were not lavish or extravagant. Rather, they were celebrations that broke up the monotony, depression, and isolation of life on the northern plains. In other words, holidays were times of joy and celebration—times for children to have childhoods. This revised narrative, the new German-Russian story, allows them to maintain their ancestral identity, an identity they inherited from their grandparents and parents, while at the same time it allows them to account for their childhood experiences.

Emmanuel Bertsch uses his memories of holiday traditions, in particular Christmas, to reconstruct his personal narrative, and as a result his ethnic group's collective identity. Growing up in the 1930s, Emmanuel remembers both sacred and secular traditions associated with the Christmas holiday. He remembers a bad character, by the name of Saint Nicholas, who all children feared. He also recalls receiving practical goodies, such as boxes of apples, oranges, or nuts. The entire Bertsch family, of course, shared the goodies. The best parts, according to Emmanuel, were church services. "Everything was reborn again" this time of year, Emmanuel insists. Midnight Mass was beautiful. In the words of Emmanuel, holidays provided "hope for tomorrow."[92]

[92] Interview with Bertsch.

CHAPTER 9

A Fun Time: Coming of Age

"I can still see it like it was today," exclaims Florence (Mitzel) Engelhart of Venturia, North Dakota. Born and reared near the small town of Strasburg, North Dakota, Florence has fond memories of her teenage years, especially her high school proms. She remembers that some were in the basement, and some were in the cafeteria of Strasburg High School, while some were at the city center. They always had a live band. As she tells it, "Ervin Boschee was the big band during my adolescence." Her father always made sure she and her sisters, Lucretia and Rita, had "nice, long dresses" for prom. She remembers wearing a light blue off-the-shoulder dress with a shawl her junior year and a beautiful pink dress her senior year. Once the girls were juniors and seniors, their father, Simon Mitzel, permitted young men to escort them to and from prom. He, however, expected their dates to have his daughters home within fifteen to twenty minutes of its conclusion. Her father's strictness never prevented Florence from enjoying her high school proms; she found the decorations, food, music, and boys unforgettable.[1]

Coming of age on the northern plains was "a fun time," in the words of Hilda Melita (Lindemann) Backfish. Yearning to share their memories of growing up German-Russian on the northern plains, Florence and Hilda participated in the Dakota Memories project. They wanted to share their happy memories—of proms, dates, and high school—as well as reaffirm the German-Russian story.[2] The teenage years of German-Russian descendants were not all that different

[1] Interview with Florence (Mitzel) Engelhart.

[2] Interviews with Florence Engelhart and Hilda Melita (Lindemann) Backfish.

from others who grew up in the rural areas of America and Canada during the early- to mid-twentieth century. Although adolescence was a newly recognized stage in child development at that time, teenagers partook in distinctly adolescent activities, like attending high school, going to dances, participating in athletics, and going on dates. For most Dakota Memories narrators, coming of age was a liberating experience, perhaps the most pleasurable aspect of growing up on the northern plains. Memories of these activities—activities that most narrators thoroughly enjoyed—have altered the established narrative of German-Russians. Their story is no longer solely a tale of hardships, isolation, and misery. The new narrative includes stories about double dating; playing organized sports, like basketball and baseball; and celebrating love. This change in the German-Russian story allows their ethnicity to flourish, as it provides these descendants with an incentive to remember.

Adolescence is a transition stage, a social construction, which exists between childhood and adulthood, and as a result is an essential category of analysis for an examination of growing up on the northern plains. In *Huck's Raft*, a work about the history of American childhood, Steven Mintz examines the emergence of adolescence. According to him, society began viewing children through new lenses as early as the mid-1700s. At that time, "[P]arents," Mintz says, "began to regard children as innocent, malleable, and fragile creatures who needed to be sheltered from contamination."[3] He acknowledges, "[C]hildhood and adolescence as biological phases of human development have always existed."[4] Yet, "[T]he way in which childhood and adolescence are conceptualized and experienced are social and cultural constructions that have changed drastically over time."[5] Mintz points out that the so-called modern concept of childhood gained popularity throughout the nineteenth century. By the twentieth century, as people became increasingly aware of the changing periods of child development, they defined new stages to match their observations.[6]

Adolescence, as Mintz says, is "a distinct, conflict-driven stage characterized by intense passions, a penchant for risk-taking, and wildly fluctuating emotions."[7] Some scholars describe adolescence using Victor Turner's anthro-

---

[3] Steven Mintz, *Huck's Raft: A History of American Childhood* (Cambridge: Belknap Press of Harvard University, 2004), 3.

[4] Mintz, *Huck's Raft*, 4.

[5] Mintz, *Huck's Raft*, 4.

[6] Mintz, *Huck's Raft*, 3-4.

[7] Mintz, *Huck's Raft*, 187.

pological findings on liminal space.[8] In "Liminality and Communitas," Turner roots his idea of liminal space in the foundational work on "rites de passage" by Arnold van Gennep.[9] Turner describes liminality as a period of transition from one state to the next, similar to rites of passage. He says liminality is rather ambiguous, fitting in neither the preceding nor the following stage, and not in itself unique from either stage. Liminality, Turner concludes, ends with the arrival into a new stage. Thus, it is not a stretch for scholars to use this concept of liminal space to explain the stage of adolescence, a period of transition between childhood and adulthood.[10]

According to Mintz and Glenn C. Altschuler, a historian who specializes in American popular culture, psychologist G. Stanley Hall was the so-called father of adolescence.[11] In 1904, Hall published *Adolescence,* popularizing the self-titled term. He argued that adolescents existed in neither the stage of childhood nor adulthood, yet were not completely isolated from either stage. He concluded that childhood needed to be prolonged, as adolescents needed age-appropriate activities to aid in their transition to adulthood, to help them in "the turbulent period between puberty and physical maturity."[12]

While adults constructed the concept of adolescence, youth created its own culture. "Adolescent culture," Mintz states, "was largely the creation of young people themselves, who created a distinctive high school culture centering on sports, dating, and the peer group."[13] Although adolescents were increasingly attending high school during the early twentieth century, not all teenagers had the ability or opportunity, especially those living in the rural regions of America and Canada. Thus, Mintz's term, "high school culture," may be misleading. It may be more appropriate to use Hanawalt's term, "youth culture," to describe the distinctive culture that emerges in rural areas throughout the 1900s.[14] This

---

[8] Victor Turner, "Liminality and Communitas," in *The Ritual Process: Structure and Anti-Structure* (Ithaca, NY: Cornell University Press, 1969), 94–130; also see Barbara A. Hanawalt, *Growing Up in Medieval London: The Experience of Childhood in History* (New York: Oxford University Press, 1993).

[9] Turner, "Liminality and Communists," 94-130.

[10] For more information, see Mintz, *Huck's Raft*; Hanawalt, *Growing up in Medieval London*; Glenn C. Altschuler, *All Shook Up: How Rock 'n' Roll Changed America*, ed. David Hackett Fischer and James M. McPherson, Pivotal Movements in American History (New York: Oxford University Press, 2003); Glen H. Elder Jr., *25th Anniversary Edition of Children of the Great Depression* (Boulder, CO: Westview Press, 1999).

[11] Altschuler, *All Shook Up*, 108; Mintz, *Huck's Raft*, 88.

[12] Mintz, *Huck's Raft*, 196.

[13] Mintz, *Huck's Raft*, 197.

[14] Hanawalt, *Growing up in Medieval London*, 11.

new youth culture consisted of chores and responsibilities, high school, sports and organized activities, jobs, dating, and weddings. Undoubtedly, this culture contributed to the formation of their identity, including their ethnic identity.

During discussions of their adolescence, narrators of the Dakota Memories tend to emphasize stories about chores and responsibilities. These stories are the ones that closely resemble and thus reaffirm their established collective memory German-Russians as being hard-working immigrants on the northern plains. They demonstrate that with age came more responsibilities, as teenagers were responsible for more labor-intensive or time-consuming chores. Since German-Russians remember having a largely agricultural life, stories recount adolescents learning to use farm machinery, like headers, binders, and tractors. Some narrators insist they took on more household responsibilities, like child rearing, meal preparation, and animal tending. A few narrators recount stories of their parents hiring them out. These narrations reaffirm the German-Russian story. By emphasizing their memories of chores and responsibilities rather than their memories of fun and enjoyment, these narrators are staying true to their ethnic identity—an identity of hard-working Germans from Russia.[15]

Simon Schwan, a Dakota Memories narrator from Michigan, North Dakota, who grew up near Rugby, North Dakota, remembers with pride laboring in the fields throughout his teenage years. By the time he was twelve, Simon was operating a two-bottom plow with five horses. He was also running the binder during harvest season. Threshing was an exciting time, Simon insists. When he describes threshing season, Simon mentions that he will never forget the smell of tar on the belts of threshing machines. He states, it "stays with you for the rest of your life." Although he thoroughly enjoyed growing up on a farm, Simon acknowledges it was hard work. When he was thirteen years old, Denis, his father, held Simon out of school for two weeks to haul grain to town. Simon remembers hauling about fifty bushels per day, totaling nearly seven hundred bushels for the two-week period. It appears that the Schwan family, in particular Simon's father, valued farm labor more than public school. In fact, Simon only completed the seventh grade before beginning his career as a full-time farmer.[16]

---

[15] Interviews with Nicholas F. Axtmann, Arthur R. Bender, Pauline "Polly" (Wock) Berger, LeRoy Buck, Stanley Graven Dobler, Florence (Mitzel) Engelhart, Clara (Ell) Ferner, Henrietta (Gall) Fiechtner, Emmanuel Gene Jerger, Adam A. Kaseman, interview; Lorraine (Meidinger) Kaseman, Eileen M. (Burlack) Ketterling, Leona (Miller) Klundt, George Kraft, Larry Kraft, interview; Peter A. Kraft, LeRoy H. Presler, Arthur E. Schmidt, Wendelyn Vetter, and Delmar D. Zimmerman.

[16] Interview with Simon Schwan.

Image 9.1. Members of the Riehl family stacking hay, date unknown. This labor-intensive task exemplified the cooperation, endurance, and pride German-Russian families invested in securing winter feed for their livestock. *Photograph from the John Riehl Collection, GRHC.*

Wilma (Beck) Haupt also remembers taking on more chores and responsibilities during her teenage years of growing up near Eureka, South Dakota. Wilma remembers getting more involved in gender appropriate tasks, such as cooking, baking, gardening, and canning, as well as taking on more responsibilities traditionally reserved for young men, like fieldwork and butchering. She reminisces about a number of her tasks—making dumplings, removing and cleaning intestines, making pickles, digging potatoes, hauling grain, and caring for children. While Wilma emphasizes her chores and responsibilities during her oral history, she never outright complains about them. She just takes care to emphasize the amount of work she faced as a teenager, thereby staying true to her German-Russian identity.[17]

Larger families occasionally hired out children to other families. Hiring out was a win-win situation, as it was one less mouth to feed for larger families,

[17] Interview with Wilma (Beck) Haupt, interview.

and one more laborer for smaller families needing assistance, either in the field or in the home.[18] Michael Jahner, from Tramping Lake, Saskatchewan, has an atypical definition of the term "hiring out." Michael grew up in a rather average size German-Russian family, with four brothers and five sisters. When he was just eighteen years old, his father passed away, leaving his mother, Francis (Lemaire) to maintain the farm. Since Michael was the youngest boy, his mother asked him to seek outside employment. She already had the older boys running the farm, and the family needed additional income. Thus, Michael remembers he spent most winters doing construction work in various regions of Saskatchewan. Michael's family members expected him to send most of his earnings home for the good of the family. Furthermore, during the summer months, they expected him back home on the farm to help with harvest.[19]

Circumstances, farm labor or lack of financial funds, sometimes prevented German-Russian children from attending high school, but not all. Mintz notes in *Huck's Raft* that high school was an essential part of the teenage culture.[20] Yet, German-Russian narrators recall finishing the eighth grade if they were lucky, and not much else in regard to a formalized education. They insist their families needed them on the farm, a sentiment that reaffirms the German-Russian story of hardships and sacrifice. Then again, there are some narrators with fond memories of attending high school. These narrators reminisce about their experience attending secular or religious schools, boarding in town or in school dormitories, and having a variety of classes as well as teachers. By sharing their memories of high school, these narrators veer away from the shared memory. Their stories reveal that the second- and third-generation Germans from Russia had opportunities on the northern plains, the type of opportunities their grandparents and parents had worked so hard to provide.[21]

German-Russian descendants fortunate enough to attend high school usually went to a secular school in a nearby town.[22] Born near Napoleon, North Da-

---

[18] Interviews with LeRoy Buck, Louis Helfrich, Michael Jahner, Klundt, LeRoy Presler, Herbert Thurn, and Schmidt.

[19] Interview with Jahner.

[20] Mintz, *Huck's Raft*, 197.

[21] Interviews with Hilda Melita (Lindemann) Backfish, Clarence Bauman, Margaret Bitz, Alice Ruth (Miller) Buck, Edna Lu Ella (Weisz) Buck, Edwin "Emil" Dewald, Florence Engelhart, Henrietta Fiechtner, Dorothy (Grasser) Miller, Julius Miller, Rose Miller, Carolyn "Carol" (Marquart) Mock, Eva (Helfrich) Mutschelknaus, Peter Nagel, Gilbert Ost, Elaine Marie (Volk) Selensky, Herbert Thurn, and Cora (Wolff) Tschaekofske.

[22] Interviews with Edna Buck, Dewald, and Florence Engelhart.

kota, Edwin "Emil" Dewald remembers attending high school in town, which was eleven miles from his family's farm. Given the distance and the potential for hazardous winter weather conditions, Emil boarded in Napoleon while attending high school. When asked to share his memories about high school, Emil reminisces about the transition to town school, mischievous behaviors, as well as the curriculum. He insists the transition was not that bad, as there were a few other farm children attending high school in town, as well. Thus, he had a built-in social network. With a small smirk on his face, Emil precedes to share his memories of creating mischief—talking too much in class and throwing spitballs at the teacher. He concludes it was not a good idea but usually entertaining. Pulling pranks and being disruptive, Emil remembers, interrupted daily lessons—lessons in spelling, geography, arithmetic, science, and law.[23]

Some Dakota Memories narrators remember their parents sending them to a religious high school, usually a boarding school, rather than a secular high school. This practice seems more common amongst Catholic Germans from Russia than Protestant Germans from Russia.[24] Born in Logan County, North Dakota, Margaret Bitz remembers growing up in a deeply religious German-Russian home. In fact, after completing the eighth grade, Margaret started attending a Catholic boarding school, Saint Frances Academy, being destined to become a nun. She describes a typical day beginning with Mass in the morning, followed by breakfast. After breakfast, Margaret recalls, all the children attended classes. She remembers that math and vocabulary were two of her favorites. Following classes, students who boarded at the school were responsible for daily chores, like peeling potatoes, dusting furniture, mopping bathrooms, and washing dishes. Margaret seems to have fond memories from her residency at this boarding school, although her destiny changed. She became an educator and counselor.[25]

Boarding in town or at school was common for German-Russians on the northern plains during the early- to mid-twentieth century. Quite a few Dakota Memories narrators, who attended high school, whether secular or religious, recall boarding near the school.[26] Edna Lu Ella (Weisz) Buck, born in rural Lehr, North Dakota, remembers boarding in town when she was attending high

[23] Interview with Dewald.
[24] Interviews with Bitz, Clara Ferner, Elaine Selensky, and Leo Peter Herauf.
[25] Interview with Bitz.
[26] Interviews with Bitz, Edna Buck, Dewald, Florence Engelhart, Helfrich, Peter Nagel, and Herbert Thurn.

school. Moving in with a "caretaker," as Edna recalls, was quite the experience for her at fourteen years of age. She remembers there were eleven other students boarding in this house at the same time. The girls slept downstairs; the boys slept upstairs. "Grandma Sara's" place, according to Edna, lacked a bathroom or an outhouse. She remembers the boarders always had to sneak into the high school, which was only one block away, to use the toilet and shower. Occasionally, the janitor caught them and sent them back to "Grandma Sara's place." Edna remembers her caretaker as a nice, yet strict, woman. Grandma Sara, as the children called her, made sure they got up in time for school and went to bed at a decent hour. She, Edna points out, never coddled the children. Her boarders were responsible for cooking and cleaning, regardless of age or gender.[27]

The Dakota Memories narratives on high school experiences vary, especially in regard to curriculum and afterschool activities. Besides the traditional three Rs, the high school curriculum varied from region to region. Participation in afterschool activities, such as basketball, baseball, pep squad, hockey, and wrestling, also varied from individual to individual.[28] Growing up in Golden Valley, North Dakota, Hilda Melita (Lindemann) Backfish remembers "[I]t was a fun time" attending Golden Valley High School. She remembers taking a variety of classes in high school, including (but not limited to): physical education, world history, grammar, literature, agriculture, driver's education, and general science. Hilda never cared for world history or grammar. Literature, however, was her favorite subject. She had a serious passion for reading. Some of her favorites were *Heidi*, *Rover Boys*, *Cinderella*, *Grapes of Wrath*, and *How Green Was My Valley*. "I liked the classics," Hilda concludes.[29]

As the Dakota Memories narrations attest, children who attended high school had more opportunities for fun. High schools had some after-school programs, ranging from organized sports to creative activities to school clubs.[30] Born and reared in Wishek, North Dakota, Marvin G. Fiechtner has fond memories of playing organized sports in high school. This self-proclaimed "town kid" remembers playing football for Wishek High School. In his heyday, Marvin played largely in the backfield, as quarterback, halfback, and fullback. He remembers they played against the "Fort Yates Indian School" quite often. They

---

[27] Interview with Edna Buck.

[28] Interviews with Backfish, Marvin G. Fiechtner, Herbert Thurn, and Tschaekofske.

[29] Interview with Backfish.

[30] Interviews with Marvin Fiechtner, Mock, and Herbert Thurn.

Image 9.2. St. Anthony's hockey team, *circa* 1927. Participation in hockey provided German-Russian youth with opportunities for community pride, social connection, and joyful recreation during harsh prairie winters. *Photograph from the Christina Rose Marie (Bast) Krismer Collection, GRHC.*

had to take a ferry across the Missouri River for their games except during the winter months, when they simply drove across the frozen river. Thinking back on those years, Marvin remembers one player on Fort Yates's team who had a reputation for being "big, fast, ugly, and mean." Marvin says, "[Y]ou ran fast if he was coming for you!"[31] These activities—organized sports—prompted integration into mainstream society, as children meet others from outside their ethnic group.

While it was socially acceptable for girls to participate in organized games in physical education classes or on a neighbor's farmstead, there were limited opportunities for them on organized high school athletic teams. As Hilda Backfish proclaims, there were "no girl teams for competition." At Golden Valley High School, where Hilda went to school, there was a Pep Club for girls.[32] Hilda remembers belonging to this club. "We didn't have uniforms," she states. This

[31] Interview with Marvin Fiechtner.

[32] Interviews with Backfish, Magdalena Miller, and Tschaekofske.

club was similar to cheerleaders, just without uniforms. They had, in Hilda's words, "organized yells." According to Hilda, they made up their own cheers. They would rendezvous wherever the boys' basketball team was playing and cheer for them. "We cheered for our guys," Hilda says, even though they rarely won. "In fact," Hilda continues, "I don't think we ever [won] when I was in high school." Nevertheless, Hilda says, their spirit was always there.[33]

Not all teenagers attending high school had the opportunities to participate in the after-school games and activities. Dakota Memories narrators attest that it was hard, if not impossible, for farm children to participate. Unless they were boarding in town and without work responsibilities, farm children were needed back home, on the farm, immediately after school.[34] Clarence Bauman of Java, South Dakota, for instance, remembers being envious of town children because they could participate in the extracurricular activities. Clarence insists that by the time he was thirteen years old his parents, in particular his father, Fred G., expected him to do "man's work." As a result, Clarence never played basketball, or any other sport for that matter, for Java High School. After school, Clarence says, he and his brother had to hitch up the horses quickly and head home, which was approximately five miles from town. Then they had to change their clothes quickly and get to work, thus demonstrating the liminality adolescent farm children experienced. Clarence was neither a child, nor an adult.[35]

Sometimes, as Dakota Memories narrators attest, parents joined forces to thwart their children's plans for after-school activities. For instance, proms and school dances were hot issues in towns. Some Protestant religions found the activity of dancing sinful, and as a result, parents were adamant about forbidding schools from hosting dances. School administrations, in other words, supported the conservative beliefs of German-Russian parents.[36] Born and reared on the outskirts of Streeter, North Dakota, during the 1930s and 1940s, Alice Ruth (Miller) Buck remembers the school board of Streeter High School getting involved in discussions about proms. She insists the school board said, "[T]here was no dancing in school," therefore, "there was no dancing in school." Her parents, both of whom were Lutherans, were completely open to the idea of dancing. However, the Streeter School Board was strict, and they forbade any

[33] Interview with Backfish.

[34] Interviews with Clarence Bauman and Peter Nagel.

[35] Interview with Clarence Bauman.

[36] Interviews with Alice Buck, Mock, and Theresa (Meier) Eissinger.

dancing at the high school. Although Alice never attended a school dance, she does remember attending plenty of barn dances. She enjoyed hearing "a good fiddler or guitarist."[37]

In addition to, or instead of, attending high school, some teenagers in early to mid-twentieth century sought employment outside the home. A number of Dakota Memories narrators reminisce about their first job, about gaining some autonomy. As childhood historian Riney-Kehrberg states, "[G]rowing up did not just mean making a significant, or even adult contribution to the family enterprise, but gaining independence and eventually leaving the family home."[38] For Dakota Memories narrators the first step in transition from adolescence to adulthood was outside employment. Several narrators reminisce about working as a retail clerk at a local store, teaching at a country school, laboring on a nearby farm, serving their country—voluntary or involuntary—in the Civilian Conservation Corps (CCC) or military, or attending college. Regardless, these positions taught adolescents a good deal of responsibility by granting them autonomy. Through these responsibilities, teenagers were learning to be adults. Yet, they were still subject to their parents' authority and influence.[39]

Although it was more common for young girls to stay on the farm, assisting with the farm labor and domestic responsibilities until marriage, a few Dakota Memories narrators reminisce about getting outside employment during their youthful years.[40] Elda (Schulz) Rasch, born in Fredonia, for instance, remembers working as a store clerk during her adolescence. It was her first experience working outside the home, and she thoroughly enjoyed it. Then again, she was still within her ethnic community, and therefore under the watchful eye of relatives and family friends.[41]

Since the majority Dakota Memories narrators grew up during the Great Depression or its aftermath, they tend to emphasize the lack of opportunities during this decade—a decade they commonly refer to as the Dirty Thirties. Several narrators, however, acknowledge the New Deal programs that offered

[37] Interview with Alice Buck.

[38] Riney-Kehrberg, *Childhood on the Farm*, 183.

[39] Interviews with Paul A. Hauck, Henrietta Fiechtner, Alvin E. Graf, John J. Gross, Mathias A. Gross, Herbert Herman, Reinhold R. Klein, Rupert Kleingartner, Clarence Meidinger, Joseph Michael Meier, Elda (Schultz) Rasch, and Tillie (Deisz) Rath.

[40] Interviews with Rasch, Mutschelknaus, Tillie Rath, Elaine Selensky, and Agatha "Peggy" Sellinger.

[41] Interview with Rasch.

Image 9.3. Elda (Schulz) Rasch, outside her first job in Fredonia, North Dakota, *circa* 1940s. Her work in town reflected new opportunities for independence and changing roles for young German-Russian women, marking a shift in community and family expectations. *Photograph from the Elda (Schulz) Rasch Collection, GRHC.*

relief and assistance, like the Civilian Conservation Corps.[42] Born in 1915 and reared in Gackle, North Dakota, Herbert Herman remembers serving in the CCC. This World War II veteran compares life in a CCC camp to life in a military barrack. Serving for the CCC in Wishek, North Dakota, and Camp Crook, South Dakota, Herbert earned thirty dollars per month, twenty-five of which the government sent directly home to his family. He recalls that upon returning home, his mother, Ottilie (Roloff), asked for his permission to use this money for an addition to the Herman home: a bathroom. The fact that his family saved this money, and asked for permission to use it, impressed young Herbert.[43]

Dakota Memories narrators also spend time reminiscing about their military service. Some narrators volunteered to serve for their countries, while the

[42] Interviews with Graf, Math Gross, Herman, Edmund Opp, Tillie Rath, and Irene (Bender) Schauer.

[43] Interview with Herman.

Image 9.4. At eighteen years old, Roy F. Nill began his service in the United States Army, serving for two years during World War II, *circa* 1944. His service reflects the deep sense of loyalty, sacrifice, and pride that German-Russian families felt as they supported their sons in defending their new homeland. *Photograph from the Roy F. Nill Collection, GRHC.*

national government drafted others. Unfortunately, this experience was traumatic for veterans, and as a result, they occasionally opt not to share their experience. By putting their lives in jeopardy these young men shortened their adolescence.[44] Rupert Kleingartner, from Gackle, North Dakota, remembers spending a year of his life serving his country. From 1945 to 1946, Rupert was part of the United States Army, and he opened up about his experiences in Okinawa, Japan, during his oral history session. Not only did Rupert see firsthand the aftermath of World War II, but he also witnessed a number of new environmental extremes, like typhoons. These typhoons, as Rupert recalls, forced

[44] Interviews with George L. Berger, Erwin John Gross, Math Gross, Paul Hauck, Laura (Schulz) Hellmuth, Herman, Reinhold Klein, Rupert Kleingartner, Meidinger, Rose Miller, Peter Nagel, Roy F. Nill, Ost, Emil Roesler, Joseph J. Schneider, and Herbert Thurn.

him, as well as other American soldiers, to seek shelter in abandoned Japanese bunkers with human remains. His time in the army was unforgettable and transformative.[45]

Logan County, North Dakota, native Erwin Geiszler also wanted to share his memories of World War II for the Dakota Memories project. The United States Army drafted Erwin as a young man. He remembers training with two-by-fours, because there was a shortage of weapons. After training, the army deployed Erwin's unit to Japan and then to the Philippines. "They introduced us to the war pretty well," he says. Erwin insists his commanders encouraged the soldiers to take no prisoners. "I never saw a prisoner until we got to . . . the Philippines." This, after all, as Erwin points out, was after the bombing of Pearl Harbor on December 7, 1941. Although still a child in the minds of his parents, John and Christine (Klein), Erwin was having adult experiences overseas as a part of the United States Army.[46]

Teaching was a common first job for German-Russian descendants, especially adolescent girls. According to Dakota Memories narrators, teachers in rural schools were not always required to have a high school education, much less a college education. They needed to pass the eighth-grade exam and obtain a teaching certificate. Thus, country schoolteachers were quite young and inexperienced. Riney-Kehrberg points out that although "graduating from the eighth grade, passing a teacher's exam, and taking a position in a local school might seem to qualify a youngster for a type of adulthood . . . parents often remained the masters of children's fate."[47] In other words, younger children may perceive adolescents as adults in the classroom, but outside of the classroom, parents continued to view adolescents as children subject to their rules.[48] This was another example of the liminality of adolescence.

Born near Wishek, North Dakota, Henrietta (Gall) Fiechtner remembers teaching school when she was just seventeen years old. Following her high school graduation, Henrietta did a three-course summer school program for a

[45] Interview with Rupert Kleingartner.
[46] Interview with Geiszler.
[47] Riney-Kehrberg, *Childhood on the Farm*, 186.
[48] Interviews with Backfish, Lorraine (Mueller) Kleingartner, and Esther Christina (Mertz) Opp.

special teaching permit. The following winter she was teaching seven to eight children at a rural school fourteen miles from her parents' place. During the winter months, she lived with the parents of some of her students, as they were closer to the school. This boarding experience was the first time Henrietta had lived away from home. She describes boarding as a good experience, as the couple she resided with were nice and generous.[49]

Many German-Russian farm children spent their adolescence laboring on their family's farm, rather than seeking outside employment. Their parents needed the labor, and they (more often than not) were destined to inherit the family farm, as Dakota Memories narrators attest. Thus, it was common for children to spend their transitional years working as a field hand or domestic worker.[50] Joseph Michael Meier, who was born and reared in Tramping Lake, Saskatchewan, remembers his years on the Meier farmstead. Once the Meier boys finished school, Joseph recalls, they simply stayed on the farm. Joseph describes his adolescent years as hard and laborious. He reminisces about the never-ending responsibilities associated with farming. Each season started, he claims, with the Meier boys spending about one week cleaning seeds. Then, they had to adjust the farm equipment, getting it ready for seeding the fields. Following the seeding process, Joseph remembers doing the summer fallowing, which he describes as a long and slow process. Then, it was harvest time. "There was always something to do," Joseph says.[51]

Only a few narrators, such as Reinhold R. Klein, from Wishek, North Dakota, have memories of going off to college. Reinhold attended North Dakota Agricultural College (now North Dakota State University) from 1941 to 1942 in Fargo. The recipient of a Sears Roebuck scholarship—$125—Reinhold was determined to be an agriculture teacher. Since tuition was $30 per quarter, he never paid for tuition or books. Yet, he needed to raise money for his boarding expenses. "I raked a lot of leaves around Fargo," he remembers. He also recalls changing a number of storm windows in downtown Fargo. For this seventeen-year-old boy, this was quite an experience. He was, after all, living, unsupervised, nearly two hundred miles away from home. Thereby, Reinhold was transitioning into adulthood at a faster pace than those who remained at home, on the farm, under the constant supervision of their parents.[52]

---

49 Interview with Henrietta Fiechtner.

50 Interviews with Clarence Bauman, Edna Buck, Meier, and Ost.

51 Interview with Meier.

52 Interview with Reinhold Klein.

At this stage in child development, teenagers not only began experimenting with adult roles in the workforce, but also in their relationships. Courting and dating were common practices for teenagers on the northern plains, as well as throughout North America. Then again, Dakota Memories narrators remain hush-hush about these practices. This silence is largely the result of two factors: embarrassment as well as shame on the part of the narrators and fear on the part of the oral historians. Dakota Memories oral historians have expressed fear in approaching controversial or sensitive topics, as they do not want to ruin their rapport with their narrators. In the early- to mid-twentieth century, people did not openly discuss sex and sexual acts, like kissing, necking, and petting. Therefore, narrators remain uncomfortable discussing these issues. Only a few narrators willingly share their experiences or make insinuations regarding their experiences, leaving it up for interpretation.[53]

When asked about their dating practices, Dakota Memories narrators reminisce about group dating, or going out with a group of friends. They insist it was quite uncommon for a boy and a girl to go somewhere like the movies or a dance alone. Instead, parents preferred their children go places in larger groups.[54] Pauline "Polly" (Wock) Berger, from Stark County, North Dakota, has fond memories of dating in the 1940s, although Polly insists they "didn't date per se." Rather, a group of teenagers would meet at a dance or movie. Polly was sixteen or seventeen years old when her parents started letting her attend these group dates. She remembers they usually had dances in an octagon building or at the Holiday Inn in Stark County. Polly points out her parents were quite strict when it came to dating. The Wock girls could only attend dances if they had completed all their chores, and as long as it never deterred from their productivity for the next day. Polly still concludes, "[I]t was good times!"[55]

Narrators, especially women narrators, tend to avoid topics of intimacy when reminiscing about dating or courting. One woman, Alma (Janke) Schott of Gackle, North Dakota, almost opens up to her oral historian. When talking about the appropriate behaviors of young girls in the 1930s and 1940s, Alma stumbles onto the topics of dating and premarital sex. Embarrassed by the subject, Alma starts to blush and giggles like a teenage girl. She even brings up her

---

[53] Interviews with Clarence Bauman, Delores (Kuhn) Becker, Polly Berger, Rosemary (Fisher) Berger, Alice Buck, Edna Buck, Dewald, Florence Engelhart, Henrietta Fiechtner, Geiszler, Selma (Job) Lapp, Dorothy Miller, Mock, and Esther Opp.

[54] Interviews with Polly Berger, Alice Buck, Florence Engelhart, Henrietta Fiechtner, and Geiszler.

[55] Interview with Polly Berger.

hands to hide her face and lowers her voice. She says her mother warned her off premarital sex. Nonetheless, Alma acknowledges that it did happen.[56]

Dakota Memories narrators who share memories of kissing, necking, and petting are largely men. Nevertheless, they only vaguely describe intimacy, or they leave their descriptions open for interpretation.[57] For instance, Clarence Bauman of Java, South Dakota, points out that the older children "had their own fun" during noon hour. With a sly smirk on his face, Clarence insinuates that teenagers were engaging in intimate relations. Yet, exactly what this entails, kissing, necking, petting, intercourse, Clarence leaves open for interpretation.[58] Erwin Geiszler's story also leaves a lot open to the imagination. When asked about the dating practices in the 1930s, Erwin responds by chuckling. He continues, "[W]e went girling [*sic*]" quite often after he returned from the CCC. This activity involved taking girls to shows, dances, or movies. One of his favorite alternatives to those activities, as Erwin points out, was parking. He enjoyed taking his dates out to a nearby haystack under the moonlight. After a long pause, Erwin begins laughing. Without any more information on this practice, Erwin leaves it up to his audience's imagination to fill in the blanks.[59]

While intimacy was not one of the favorite topics to reminisce about during their oral history sessions, Dakota Memories narrators enjoy sharing stories about meeting their spouses. Some narrators insist their spouse was the only person they ever dated, or at least ever seriously dated.[60] Alice Buck started dating in 1948, at sixteen years old. Unlike most female narrators, Alice is quite candid about her dating experiences, reminiscing about "smooching" in some dark places and dating men other than her future husband, LeRoy. Nevertheless, she does take pride in sharing the story about falling in love with LeRoy. She first met him at a New Year's Eve dance in Gackle, while he was on a date with another girl. The next day, New Year's Day, there was a dance in Streeter. She was planning to go to the dance with her girlfriend, Edna Schaffer. Unbeknownst to Alice, Edna's boyfriend, Leo Kinsley, was good friends with the handsome boy Alice had met the night before, LeRoy. When the boys arrived at Edna's house, Alice's life changed forever.[61]

---

[56] Interview with Alma (Janke) Schott.
[57] Interviews with Clarence Bauman, Alice Buck, Geiszler, Joseph C. Leier, and Schott.
[58] Interview with Clarence Bauman.
[59] Interview with Geiszler.
[60] Interview with Alice Buck, Lapp, Rosa (Schneider) Laturnus, Mock, Mutschelknaus, Armelia "Millie" (Ibach) Nagel, and Ost.
[61] Interview with Alice Buck.

Image 9.5. Irene (Bender) and Elmer Schauer married on November 27, 1949. Their wedding symbolized family continuity and community pride, marking an important rite of passage for German-Russian families. *Photograph from the Irene (Bender) Schauer Collection, GRHC.*

Falling in love and marrying typically marked the end of adolescence for German-Russian descendants. As anthropologist Victor Turner acknowledges, traditionally a ceremony or ritual marks the transition from liminality into a new stage, such as a wedding marking the transition from adolescence to adulthood.[62] Although Dakota Memories narrators reminisce about teaching in a country school, serving in the United States or Canadian military, and working for the CCC, they do not view themselves as adults until entering into a union of marriage. For narrators, their wedding day was one of the happiest days in their lives. Unlike their ancestors—the first-generation Germans from Russia, who typically experienced arranged marriages—the second- and third-generation

[62] Turner, "Liminality and Communitas," 94–130.

German-Russians married for love, and by their own accord and usually within their own ethnic group. Thus, their narrations document joyous celebrations honoring this rite of passage into adulthood.[63]

Wedding traditions varied from religion to religion, from region to region, and from family to family. As Dakota Memories narrators attest, the age of marriage ranged from seventeen or eighteen years old to twenty-four and twenty-five years old and sometimes older. Some weddings were formal celebrations of love, taking place in churches. Grand receptions typically followed the formal ceremony, with food, drinks, and music. Other weddings were simple ceremonies of love, taking place in family homes. Small, quaint receptions usually followed the informal ceremonies with food, drinks, and conversations. Regardless, wedding ceremonies were typically joyous occasions for Dakota Memories narrators.[64] Henrietta Fiechtner, for instance, remembers getting married at the age of eighteen. According to Henrietta, Clarence proposed to her after two years of dating. She eagerly agreed. Henrietta acknowledges their engagement was short, as they married two or three months after Clarence's proposal. Thinking back, Henrietta says she had a small, simple wedding in the church parsonage. After the ceremony, their wedding guests joined the happy couple for a simple afternoon meal and some wedding cake, all prepared by Henrietta's mother, Hertha (Ketterling).[65]

For Catholic German-Russian descendants, the wedding ceremony was much more formal and long. Catholic Dakota Memories narrators insist their weddings were an all-day affair.[66] Delores (Kuhn) Becker and Rosemary (Fisher) Berger, both of whom are Catholic, remember their weddings as quite elaborate affairs. Married in 1954 and 1948 (respectively), at the ages of twenty-one and nineteen, Delores and Rosemary had very similar ceremonies.[67] Delores remembers marrying her husband, John G. (also a Dakota Memories narrator), at Saint Boniface Catholic Church at ten o'clock in the morning on January 4, 1954. She remembers there was a large dinner of chicken noodle soup served at noon, following the wedding ceremony. An afternoon of dancing followed

---

[63] Interviews with Becker, Rosemary Berger, Henrietta Fiechtner, Lapp, Barbara Leier, Julius Miller, and Mutschelknaus.

[64] Interviews with Becker, Rosemary Berger, Helfrich, and Henrietta Fiechtner.

[65] Interview with Henrietta Fiechtner.

[66] Interviews with Becker, Rosemary Berger, Polly Berger, Magdalena Hauck, Herauf, Armelia Nagel, and Elaine Selensky.

[67] Interviews with Becker and Rosemary Berger.

the noon meal. Then, they had supper in the evening, followed by more dancing.[68] Rosemary remembers her wedding also started at ten o'clock in the morning. She married George L. (also a Dakota Memories narrator) on November 4, 1948, at Saint Joseph's Catholic Church. Following the ceremony, Rosemary and George's guests dined on chicken, potatoes, cakes, and pies. After their fill of food, the wedding party went to the photography studio for their wedding photographs. Then, they headed back to the community hall to feast on supper with their guests, and dance the night away.[69] Both Delores and Rosemary recall the wedding celebrations lasted from ten o'clock in the morning until (at least) midnight.[70]

Coming of age, as Dakota Memories narrators attest, was a fun time. Although it was a period of physical and emotional change, i.e., puberty and liminality, narrators have fond memories of this stage of development. It was a liberating experience. Narrators enjoyed gaining more autonomy as teenagers, with increasing responsibilities and freedoms. During their oral history sessions, narrators reminisce about taking on new, more labor-intensive or time-consuming chores and responsibilities, attending high school and participating in extracurricular activities, getting jobs (serving in the military or CCC, teaching at a country school, or farming with the family, etc.), going on dates, and falling in love and getting married. Memories of these activities, for the most part, make narrators smile.

Understanding this period of adolescence is crucial to understanding childhood, as well as identity construction. Unfortunately, childhood historians have largely overlooked this stage of development, either by completely ignoring it or by lumping it together under an analysis of childhood.[71] Only recently have childhood historians begun examining adolescence as a separate stage in child development. For instance, Mintz's *Huck's Raft* and Riney-Kehrberg's *Childhood on the Farm* have sections solely devoted to discussions on adolescence.[72]

---

[68] Interview with Becker.

[69] Interview with Rosemary Berger.

[70] Interviews with Rosemary Berger, and Becker.

[71] For more information, see West, *Growing up with the Country;* Holt, *Children of the Western Plains.*

[72] For more information, see Mintz, *Huck's Raft*; Riney-Kehrberg, *Childhood on the Farm.*

Given that this stage of development mirrors certain aspects of childhood, yet is similar to adulthood, it is only fitting that scholars examine this transitional stage separately when discussing child development. Adolescence exists in liminal space, belonging to neither the preceding nor the following stage, but not completely separate from either.[73] Thus, an examination of this stage may complicate the narrative, but it makes the subject much more intriguing.

Using oral history to understand this stage of child development is revealing. Because people view these years as their heyday, there exists a desire to share stories from this stage of development. Furthermore, as Schacter points out in *Searching for Memory*, the closer one is to the event, the more detail (memory coding) is retained.[74] Thus, narrators tend to share more information about their teenage years than about their years in childhood, early childhood, or infancy. It is also during this stage of development that children sometimes engage in questionable activities (or display socially unacceptable behaviors). As a result, sometimes there is a reluctance to share memories about embarrassing topics, like premarital sex. Nevertheless, silence or insinuations reveal as much as, or more than, a complete narrative.[75]

Oral histories also expose varying degrees of adolescent liminality. According to the Dakota Memories narratives, after confirmation and before marriage children existed in a transitional stage—a period of liminality. They were neither children nor adults. These narratives reveal three degrees of liminality, however: substantial, standard, and marginal. Those who labored on their families' farms, at the expense of receiving a high school education, experienced substantial liminality. Those who left their families' farms to attend high school, teach country school, go to college, or serve in the military experienced marginal liminality. Those who faced a little of both—freedom during the day and family responsibilities in the evening—experienced standard liminality. As the Dakota Memories narratives show, the degree of liminality affects their memories of this transitional stage. Those who experienced only marginal liminality complain much less than those who experienced substantial liminality.

Regardless, narrators spend much time reminiscing about chores and responsibilities, high school and extracurricular activities, employment opportu-

---

[73] For more information on liminal space, read Turner, "Liminality and Communitas."

[74] For more information, see Schacter, *Searching for Memory.*

[75] For more information, see Naomi Rosh White, "Marking Absences: Holocaust Testimony and History, in *The Oral History Reader,* ed. Robert Perks and Alistair Thomson (New York: Routledge, 2004); Langer, *Holocaust Testimonies.*

nities, dating practices, and wedding traditions. Their memories of this transitional stage, for the most part, are positive. Narrators, in fact, express pride and satisfaction in their stories about adolescence and coming of age. These narrations are another example of the second- and third-generation German-Russians changing the German-Russian story. Their story is no longer so simplistic or a one-dimensional story of hardships and persecutions. The story instead is a multi-dimensional narrative rooted in an ancestral past of turmoil and a childhood past of opportunities.

Erwin Geiszler is one Dakota Memories narrator who has fond adolescent memories. Erwin was a teenager during the Dirty Thirties. During his oral history Erwin reminisces about working in the CCC camps, "girling [*sic*]," and getting married. His memories reveal that Erwin thought life was good during his adolescent years. It was at this stage in life that Erwin met his future wife, Eileen (Schlenker). They met because Eileen was boarding at his aunt's house in Gackle while she attended high school. They only started dating after Erwin served in World War II. Within a year of his returning home, Erwin and Eileen married in a double-wedding. According to Erwin, he and his brother were both getting married that year, and it was cheaper to get married at the same time. Therefore, they opted for a double-wedding. They also decided on a rather small, intimate wedding, as their father, John, was bedridden at the time. On February 23, 1947, Erwin and Eileen wed at the home of John and Christine (Klein), along with Erwin's brother and his bride. A small dinner followed the modest ceremony that formally recognized Erwin's transition into adulthood. He concludes, "It was nice."[76]

[76] Erwin Geiszler, interview.

CONCLUSION

"We Didn't Know Any Better"

"You didn't dilly-dally too long in the wintertime," exclaims Raymond Boechler, who grew up in the Allan, Saskatchewan, region. Raymond never had indoor plumbing. He remembers the outhouse was located about twenty-five to fifty meters from his parents' house. Therefore, in the winter months, "You'd run out there, you did your business, and you'd run back," he says, laughing. Furthermore, "There was no money spent on toilet paper." Raymond remembers they used old Eaton's catalogs, newspapers, scrap papers, or "whatever was available." He says, "Oh yes, that's the way it was in those days . . . we didn't complain about it." Raymond insists, "We didn't know any better."[1]

German-Russian descendants on the northern plains, like Raymond Boechler, align their narratives with the established German-Russian story, a tale of hardship and oppression. Hence, Raymond Boechler's seemingly simple statement: "We didn't know any better."[2] Yet, after a few hours of interviewing, it is apparent that Raymond's story is complex. His narrative, while rooted in the stories of his ancestors, also incorporates his childhood experiences, like riding bicycles and playing hockey. The same is true for other narrators of the Dakota Memories project. By reconstructing their ethnic identity using their childhood memories, narrators reveal two separate spheres of memory: one for their ethnic group's collective memory, with tales of hardship and oppression, and one for their individual childhood memories, with stories of fun and enjoyment. The spheres simultaneously exist, but completely independent of each

[1] Interview with Raymond Boechler.

[2] Interview with Raymond Boechler.

other, at least until the oral history process. The Dakota Memories narrations reveal that while rooted in an ancestral past, childhood experiences as well as current situations shape the new German-Russian story.

During the Dakota Memories interviews, narrators discuss a variety of topics, including but not limited to the environment, family relationships, chores and responsibilities, grade school, pastimes and activities, religious devotion, holidays and celebrations, and coming of age. Through their discussion of these topics, German-Russian memory-makers are reshaping their ethnic group's history. These narratives are bitter-sweet, revealing the tragic stories of oppression and hardships and the joyous stories of liberation and happiness.

Participants in the Dakota Memories project are fully aware that they inherited their ethnic identity from their ancestors. They pride themselves on maintaining the collective story and traditions that their great-grandparents and grandparents handed down, as well as the morals and convictions they learned in church regardless of the denomination. German-Russians of the second and third generation, at least those who participated in the Dakota Memories, insist they learned to be German-Russian through their family relationships, religious devotion, and holiday traditions. They learned about sacrifice, God, and thriftiness. Their memories of family, religion, and holidays reveal new experiences outside the established German-Russian story. Narrators remember whisker rubs, confirmation celebrations, and Christmas goodies. Thus, as narrators reveal, growing up German-Russian on the northern plains was neither all bad nor all good. This revised story is sensuous and heartfelt. It is a more humanistic story, making it accessible to future generations.

Dakota Memories narrators invoke their memories of the environment and chores to reaffirm their ethnic group's identity. These are the easiest topics to display hardships and suffering. Narrators reminisce about the subzero temperatures, blizzards, grasshopper plagues, and dust storms. They insist they had never-ending chores and responsibilities, but after time and deeper inquiry, oral historians tap into the positive memories of the environment and chores. Narrators remember roaming fields, picking prairie crocuses, and trapping gophers. They also recall turning their chores into fun: squirting milk at cats, playing make-believe, and racing horses. Sometimes narrators surprise them-

Image 10.1. Three of the eighteen Presler children, *circa* 1934. Large families were a source of pride and reflected the resilience and cooperative spirit of German-Russian life on the northern plains. Left to right: Cora Presler, Walter Presler, and Dakota Memories participant LeRoy H. Presler. *Photograph from the LeRoy H. Presler Collection, GRHC.*

selves by sharing all these happy memories. They never expected to share these memories with their oral historians.

Their stories begin to change once oral historians pose questions about grade school, pastimes, and coming of age. Memories of these activities contradict the established German-Russian collective memory—a memory that the second and third generations inherited from their ancestors. They share stories of learning a new language, playing Fox and Geese, and falling in love. Nonetheless, their pull to reaffirm traditional narrative remains. Narrators conclude these happy stories with memories of sacrifice and misfortunes. They reminisce about traveling through blizzards, lacking medical care, and sacrificing high school. Thus, this new German-Russian story is just a revised version of the ethnic group's established narrative.

This desire of German-Russians to reaffirm their ancestors' narrative at the expense of their own demonstrates the importance of an outsider examining this ethnic group. For years, German-Russians have been telling their own stories. While this perspective is fascinating, it is also self-affirming. Works by German-Russian scholars from within the ethnic group, such as Timothy J. Kloberdanz, Gordon L. Iseminger, and Ronald Vossler, perpetuate this ethnic group's established narrative, and as a result its identity.[3] German-Russians who opt to participate in oral histories, for the most part, also set out to preserve their ethnic group's story, the story rooted in tradition and heritage.[4]

Dakota Memories narratives reveal the desires of second-and third-generation Germans from Russia to maintain their ethnic group's identity—an identity of survivors. Initial narratives and responses are stereotypical. James Engelhart of Venturia, North Dakota, for example, reminisces about the difficulties his family experienced getting to and from church during spring floods in the 1940s. "It was really hard to get to" confirmation, James recalls.[5] Louis Helfrich of Glen Ullin, North Dakota, also emphasizes the hardships he experienced in the 1930s. Cleaning the barn, he says, "That was a chore!" Louis recalls cleaning

---

[3] For more information, consult Kloberdanz, "Volksdeutsche The Eastern European Germans"; Kloberdanz, "The Germans from Russia"; Kloberdanz, "The Traditions We Share"; Kloberdanz, "The Tradition of Brauche"; Gordon L. Iseminger, "The McIntosh County German-Russians: The Fifty Years," *Journal of the American Historical Society of Germans from Russia* 11 (1988); Iseminger, "'Hunhuskers,' 'Red Cross Roosters,' and 'Uncle Sam Whiskers': The McIntosh County German-Russians in World War I," *Germans from Russia Heritage Review* 27 (December 1997); Iseminger, "Are We Germans, or Russians, or Americans?: The McIntosh County German Russians during World War I," *North Dakota History* 59 (Spring 1992); Ronald Vossler, "Prairie *Spass*: The Folk Humor of North Dakota's Germans from Russia," *North Dakota History* 66, nos. 3 and 4 (Summer/Fall 1999); Vossler, *Dakota Kraut: A Memoir* (Fargo, ND: Germans from Russia Heritage Collection, 2003).

[4] Thomas Welk, interview by Bob Dambach, Linton, ND, 17 August 2001; Joseph Senger, interview by Bob Dambach, Karlsruhe, ND, 17 August 2001; Annie (Roesch) Larson, interview by Bob Dambach, Aberdeen, SD, 13 August 1999; Barbara (Schneider) Risling, interview by Bob Dambach, Scott, SK; 19 July 2000; Pauline (Neher) Diede, interview by Bob Dambach, Hebron, ND, 22 August 2002; Emma Fischer, interview by Ronald Vossler, Ashley, ND 2 July 1997; Ferdinand Vossler, interview by Leland Elhard, Wishek, ND 2003; Anna Symoniak, interview by Leland Elhard, Wishek, ND, 2003; Albert Saylor, interview by Leland Elhard, Wishek, ND, 2003: Eva Ketterling, interview by Leland Elhard, Wishek, ND, 2003; Reinhardt Hecker, interview by Michael M. Miller, Bismarck, ND, 8 November 1993.

[5] Interview with James Engelhart.

the barn every morning before going to school. It was a long and dirty process.[6] Responses about the hardships, such as James's and Louis's, are set and casual. They do not explore the complexities of the situations. For instance, both James and Louis initially sidestepped stories about confirmation and grade school experiences in order to emphasize the transportation difficulties and endless chores.[7]

German-Russians seem not to recognize the complexities of their own story. There is more to their narrative than oppression, sacrifice, and survival. While Dakota Memories narrators are determined to preserve their ethnic identity, hence their participation in an oral history project, most narrators have yet to embrace the complexities of their story. Narrators continue to perpetuate their ancestors' narrative at the expense of their own.

Many Dakota Memories narrators use the complex statement, "We didn't know any better," or something similar, during their oral histories. Raymond Boechler, for instance, uses those exact words when he recounts his outhouse experiences with me on July 21, 2006.[8] Other examples include but are not limited to Julius Miller of Dickinson, North Dakota, telling oral historian Cassie Ptacek, "We didn't know any different," and Mary Sellinger telling me, "You [ubiquitous] didn't know any better."[9] By invoking this saying, German-Russians of the second and third generations are claiming naiveté. On the one hand, they are upholding their traditional narrative by sharing stories of hard times; on the other hand, they are telling a new story by acknowledging their accomplishments. It is after all a new story, or at least, a new body of stories that traditional narrative cannot subsume.

By the end of their oral history sessions, narrators are surprised by the memories they reveal. For instance, Mildred Wald reminisces about her lack of material goods, such as toys and fashion. Growing up in the early 1940s, Mildred "dreamed of owning a pair of sandals." Initially, she focuses on the unfulfilled dream, stating her "parents never had extra money for sandals," shoes they believed to be impractical for a farm girl. Then, she begins to share her memories of making her own from cardboard—tracing her feet, cutting out the tracings, puncturing holes in them, tying them to her feet. She remembers

[6] Interview with Louis Helfrich.

[7] Interviews with Helfrich and James Engelhart.

[8] Interview with Raymond Boechler.

[9] Interviews with Julius Miller and Mary Sellinger.

walking to the end of her driveway with them before the holes gave way. Although her homemade sandals never lasted, Mildred remembers having fun trying to make them.[10]

Stories about childhood fun and enjoyment, in spite of the hardships and suffering, allow German-Russians of the second and third generation to treasure their identity. The resilience of ethnicity, in fact, lies in the creation of new experiences and new memories. By being flexible and allowing their narrative to evolve, German-Russians have prolonged their identity. Without this flex, their identity would become irrelevant. Therefore, by reminiscing about more than the blizzards, dust storms, isolation, and chores German-Russians are not forgetting their roots. Rather, they are revealing their adaptability, as well as their desire to persevere. Although Dakota Memories narrators are usually unaware of these two diverging narratives, during the oral histories they start consolidating them, thereby creating the new, revised German-Russian story and identity. The oral historians present during the interview process confirm the old German-Russian story for narrators, while, at the same time, encourage narrators to share their new story. Thus, the oral historians help these memory-makers reconstruct their narrative and play an active role in the revelation of this new story and the reshaping of the German-Russian ethnic identity.

Using oral narratives as a window into the history, identity, and memory of German-Russians on the northern plains reveals the values of this time-honored methodology. Employing the best techniques and approaches recommended by previous oral historians such as Paul Thompson and Michael Frisch, Dakota Memories oral historians created a rich collection of valuable and revealing oral narratives. These oral historians demonstrated that Thompson's life-story approach is one of most effective methods for breaking past the collective narrative and into the individual, deep memories. In addition, these oral historians revealed that using the established memory triggers are also effective tools for breaking past memory barriers. They discovered that Michael Frisch's idea of sharing authority is quite complex. These oral historians had to share their authority with not only the videographer and narrator, but also the German-Russian community as whole. Thus, it took a great deal of collaboration before the new narrative emerged.[11]

---

[10] Interview with Mildred (Glatt) Wald.

[11] For more information on effective oral history techniques and approaches, see Thompson, *The Voice of the Past;* Dunaway and Baum, *Oral History*; Frisch, *A Shared Authority*, Perks and Thomson, *The Oral History Reader.*

To tap into the deep memories, Dakota Memories oral historians established a solid rapport with their narrators, devoted time to the collection process, and were persistent in their investigations. Trust existed between the narrators and oral history teams. Narrators believed their oral historians were going to tell their story with accuracy and truthfulness. Without this trust, the oral historians would not have tapped into the deep memories (heck, they might not have even gotten *at* the collective memories). To establish this rapport, the oral historians communicated with all potential narrators in advance of the scheduled interviews. In all communications, they were transparent concerning the process and purpose. In addition, the oral history teams created safe and comfortable environments for the interviews. They held most of the interviews in the narrators' homes, where they would be most comfortable. These oral history teams also mastered the skill of patience and persistence. Relationship building takes time. These oral historians discovered that narrators only reveal the deeper memories hours into the process. Because, narrators prefer to share only stories that reaffirm their group's collective memory, these oral historians had to be patient and persistent.

The oral history process results in discoveries not only for oral historians, but also for narrators. This interactive process reveals the tension that exists between the established collective memory and the personal experiences of narrators (at least it did for German-Russians who participated in the Dakota Memories). Thus, narrators left this experience with a fuller, more complex story. With nearly two hundred oral histories in the collection, scholars have an opportunity to acknowledge an even greater complexity—a new, revised narrative for German-Russians on the northern plains. Although this new German-Russian story is not established, there are multiple suggestions of its existence throughout the narratives. Thus, at least for Dakota Memories narrators, they are creating a new German-Russian identity, rooted in their ancestral past and childhood past, yet shaped by their present conditions.

Along with the evolution of the German-Russian identity, these oral histories reveal that people cannot pickle their ethnicity. Ethnic identity will not survive if it remains immersed in brine as everything around it changes. Ethnic groups, like German-Russians, which make accommodations for change, will outlast those groups choosing to live in the past. In other words, change is essential for ethnic resilience. Ethnic groups, such as Germans from Russia, must

Image 10.2. Frank and Anna (Freilich) Hoffart before they were married on October 17, 1946. This photograph captures the joy and pride of courtship, reflecting family and community traditions that celebrated the transition into married life. *Photograph from the Frank and Anna (Freilich) Hoffart Collection, GRHC.*

incorporate new traditions into their ethnic identity, often through the process called cultural diffusion, in order to maintain its resilience.

The oral narratives also provide new insights into childhood memories and history, like the value of examining ethnicity in childhood in the American West. Much of this regional historiography of rural childhood ignores or sidesteps ethnicity (with the exception of the bountiful material emerging on Native American boarding schools). The Dakota Memories oral histories, however, show that there is a great deal to learn by examining childhood through ethnic lenses. It reveals stories about holidays and celebrations, religious practices, family heritage, and educational experiences. Since cultural customs vary between ethnic groups, regional works on childhood in the American West tend to overlook them. Yet, as the oral histories reveal, these customs played a significant role in the childhood experiences, memories, and identities of children.

The nearly two hundred memory-makers of the Dakota Memories project have personally demonstrated the value of the oral history methodology for reconstructing collective and individual memory. For the most part, this self-selected group of German-Russians begins its oral histories rooted with tales of hardships, just as their ancestors did years ago. However, by allowing trained oral historians into their homes, and eagerly participating in the entire oral history process, these memory-makers are choosing to reconstruct their narrative. This collection reveals that, although German-Russians of the second and third generation root their story in their ancestral past, their childhood experiences and memories—as well as present conditions—influence it. Thus, this new German-Russian story is more complex than originally believed. It is a story of generational change and ethnic resilience.

Given that no other regional ethnic group has created a similar body of research material, it is impossible to compare and contrast the reconstruction of ethnic identity of Germans from Russia, as explained in this work, with any other ethnic group on the northern plains. An equivalent body of material such as an oral history project with the second and third generations does not exist for any other ethnic group in this region. Scholars could and should start to investigate the reconstruction of memory and identity with the descendants other ethnic groups on the northern plains.

Nevertheless, there is a wealth of information available on growing up in the American West, which allows for a comparison of childhood experiences and ideas in the West. Using the works of Elliott West, Pamela Riney-Kehrberg, and Marilyn Irvin Holt, for instance, it is evident that German-Russian children experienced childhoods similar to most children growing up in rural areas of the West. Within their respective works, each childhood historian discusses children interacting with the environment, laboring on farms, attending one-room schoolhouses, playing make-believe, and creating mischief.[12] Dakota Memories narrators, as recounted throughout this work, have engaged in all of these activities. They, however, tell their story through the lens of their

[12] For more information, see West, *Growing Up with the Country*; Riney-Kehrberg, *Childhood on the Farm*; Holt, *Children of the Western Plains*.

ethnic identity—sharing stories about language, foodways, cultural traditions, and religion. Hence, it is important to examine ethnicity in childhood in the American West.

In the end, this work is not a study of external relations or external comparisons. Rather, it is an examination of the internal dynamics of one ethnic group with recognized agency, Germans from Russia on the northern plains. This work is concerned with the memories and identity of this particular ethnic group. This focused study examines the stories and memories of the second and third generations, their reconstruction of identity, and the emergence of a new German-Russian story. The narrations of the Dakota Memories are at the heart of this book, as they reveal the active role of German-Russian memory-makers in reconstructing the German-Russian story—and, as a result, reshaping the ethnic group's identity.

Born and reared near Dickinson, North Dakota, Eleanor (Hutmacher) Urlacher, closed her oral history with a sentiment echoed by many others. When her oral historian asked "Eleanor, why do you think it's important to tell your life story?" Eleanor responds without hesitation: for the next generation. Eleanor recalls listening to her father's stories, certain she would never forget them. But time, as it does, has blurred the details. "I didn't think I'd ever forget," she admits, "but I have." Her voice carries the weight of loss, not just of memory, but of a history that once seemed unshakable. "I thought he'd be there forever, [but] that's not the way life is."[13]

Now, fifteen years since the Dakota Memories project was completed, many of these narrators—storytellers—have long since passed. Their voices, once full of recollections and laughter, are now echoes in the archives. Their stories survive only because they were recorded. Eleanor's regret is a reminder that history is fragile—not in books or documents, but in the lived experiences passed down through generations. And once a voice is lost, so too is a piece of the past.

[13] Interview with Eleanor (Hutmacher) Urlacher.

Acknowledgments

Rooted in the oral histories and archival materials of the Dakota Memories Oral History Project (Dakota Memories), this manuscript is a product of the support and generosity of organizations, institutions, and individuals. One of the most instrumental individuals in establishing and maintaining this project was Michael M. Miller, Emeritus Director of the Germans from Russia Heritage Collection (GRHC) at North Dakota State University (NDSU) Libraries. He assisted on every level of the Dakota Memories, from drafting interview questions, to suggesting potential narrators, to securing funding. Miller was also responsible for continuing the Theresa Mack Germans from Russia History Doctoral Fellowship as the Germans from Russia History Doctoral Fellowship. He, in collaboration with Tom Isern (University Distinguished Professor and Professor of History, NDSU), established this fellowship to fund a doctoral student's research. Hence, my involvement with the GRHC's oral history project.

While I was designing, launching, and coordinating Dakota Memories, a number of internal and external organizations as well as institutions provided funding for this project. At NDSU: the Center for Community Vitality; the Center for Heritage Renewal; the College of Arts, Humanities, and Social Sciences (now the College of Arts and Sciences); the Department of History, Philosophy, and Religious Studies; the Development Foundation; the Extension Service; the Germans from Russia Heritage Collection; the Graduate School; the NDSU Libraries; and the Vice President for Research, Creative Activities, and Technology Transfer, sponsored the Dakota Memories. The Germans from Russia Cultural Preservation Foundation, Glueckstal Colonies Research Association, Canadian Embassy, North Dakota Humanities Council, and Prairie Public also provided funding at different stages and intervals of this project.

Other individuals need to be acknowledged for their financial contributions to the Dakota Memories. For example, Patricia (Gabriel) Belknap, Alice (Walter) Brossart, Julie (Opp) Burgum, Calvin K. Fercho, Arthur E. Flegel, LaVern Freeh, Bernadine (Lang) Kuhn, Roger Kuhn, Vernon Mathern, Michael M. Miller, Herb Poppke, Ila Rae Reich, and Theresa (Mack) Wald all provided significant financial support. More than one hundred other individuals also contributed to this project.[1]

Narrators of the Dakota Memories also need to be recognized for their contributions. Nearly two hundred individuals participated in this project from 2005 to 2009. These German-Russians of the second and third generations recounted their childhood experiences with oral history teams. They permitted our teams to spend hours, sometimes days, in their living rooms and kitchens. They shared their memories, talents, photographs, and documents with our teams. They took our oral historians and videographers on tours of rural landmarks and cemeteries. They treated our teams to wonderful music and delicious German-Russian delicacies. Without these memory-makers, this research would not have been possible. Please return to the Preface of this work for a complete listing of the Dakota Memories participants.

The Dakota Memories staff and volunteers also need to be acknowledged for their contributions to this project. Michael M. Miller, Acacia (Jonas) Stuckle, and I worked endlessly to keep the Dakota Memories office staffed with strong and productive individuals. The support staff were largely NDSU undergraduate and graduate students. These students spent hours conducting oral histories, processing the archival material, and designing our project website. The following individuals made significant contributions to this project: William H. Clark, Sarah (Lacher) Crary, David Mart, Eric Mischnick, Andrea Mott Glessner, Cassie Ptacek, Neall Pogue, Kendra Thompson, and Jen Wilkie. Although most of the Dakota Memories staff and volunteers were NDSU students, we did have a few professionals and non-students that need to be recognized for their contributions, as well. Michael Boechler, Bob Dambach, Mathias Hoffmann, and Tom Isern, also supported this project in various capacities.

I have also received feedback from conference commentators and attendees by presenting my research at regional and national conferences. North Dakota State University's College of Arts, Humanities, and Social Sciences (now

[1] For a list of donors, contact the Germans from Russia Heritage Collection at North Dakota State University.

the College of Arts and Sciences); Department of History, Philosophy, and Religious Studies (now the School of Humanities); Office of Research, Creative Activities, and Technology Transfer's Sponsored Programs Administration; and Tom Isern made attending these conferences possible. The feedback received at each of these conferences aided immensely in the construction of the chapters that make up this work.

I also need to acknowledge the contributions of my doctoral advisor Tom Isern. This work started out as my doctoral dissertation, and, as such, Isern has read and reread all chapters (as well as related conference papers and articles). I truly appreciate his constructive feedback. He has encouraged and supported my career in the academy, providing me with chances to share my research and grow as a scholar. I am indebted to him for giving me wonderful opportunities during the course of my doctoral program, as well as for pushing me to make it more than a dissertation.

Transitioning this study from a dissertation to a book was quite the process, and there are a few others who deserve recognition for their assistance. Melia Dayley, Western Wyoming Community College's 2016 Outstanding Graduate, generously volunteered to read chapters for clarity and readability while she was still completing her undergraduate studies. Taking on such a time-consuming task at that stage speaks to her dedication and sharp analytical skills. Now a Master of Public Administration, she is doing incredible work in Utah.

I also extend my gratitude to Suzzanne Kelley, my editor at NDSU Press, for making the editing process an enriching and rewarding experience. Publishing this work with NDSU Press is particularly meaningful given my deep commitment to education. As a teaching press, NDSU Press provided me with the opportunity to work with an exceptional team of student editors who brought fresh insights and thoughtful engagement to the manuscript. My deep thanks to Ella Hlucny, Monika Sauer, Brandon Stariha, and Breanna Walburn, who each contributed their talents to this project. Their enthusiasm for the work—expressed in their own words—was inspiring, and their diligence has strengthened this book in countless ways. Throughout this process, Suzzanne ensured careful oversight, following up behind the students while preserving the integrity of the teaching experience. It has been an extraordinary process, one that has reinforced the collaborative and educational spirit at the heart of this publication.

Finally, I acknowledge my husband and daughters—Dr. William H. Clark, Ocean Maria Clark, and Rain Lily Clark—whose unwavering love and support have carried me through this journey. Will has been my partner in every sense of the word, dedicating his time, energy, and talent to help bring this project to life. For the first three years of the Dakota Memories project, he stood beside me as the lead videographer, traveling throughout North Dakota, South Dakota, and Saskatchewan, capturing stories that became the foundation of this book. He set aside his own doctoral research to help me pursue mine, a sacrifice that speaks volumes about his generosity and belief in my work. He has been my editor, my sounding board, and my greatest motivator—offering sharp insights and steady encouragement when I needed them most. And to Ocean and Rain, my brilliant, creative daughters—you have been my constant inspiration. Your patience, understanding, and little moments of joy reminded me why preserving these histories matters. The late nights, the long hours, the endless revisions—all of it was made possible because of the love and strength I found in our family. This book is as much yours as it is mine. Thank you, from the depths of my heart.

BIBLIOGRAPHY

PRIMARY SOURCES

Dakota Memories Oral History Project

See Narrators section in the Preface for a complete list of narrators. The material collected for each narrator includes a video-recorded oral history, archival documents (photographs, family histories, immigration papers, religious certificates, and report cards), media sources (musical performances and interviews), and more. All materials associated with this project are housed at North Dakota State University, Libraries, Germans from Russia Heritage Collection, Fargo, North Dakota (ndsu.edu/grhc/dakotamemories.)

Family Histories

Neu, Leona. "Family History." Fargo, ND: Germans from Russia Heritage Collection, North Dakota State University.

Smith, Margaret (Tuchscher). "The Geiszler Brother: German-Russian-American Family History." Fargo, ND: Germans from Russia Heritage Collection, North Dakota State University, 2006.

_________. "A Tuchscher Family History." Fargo, ND: Germans from Russia Heritage Collection, North Dakota State University, 2006.

Memoirs

Clark, Dolly Holliday. "Memoir of a Country Schoolteacher: Dolly Holiday Meets the Ethnic West, 1919–1920." *North Dakota History: Journal of the Northern Plains* 59, no. 1 (Winter 1992): 30–45.

Diede, Pauline Neher. *Homesteading on the Knife River Prairies.* Edited by Elizabeth Hampsten. Bismarck, ND: Germans from Russia Heritage Society, 1983.

Fields, Mamie Garvin and Karen Fields. *Lemon Swamp and Other Places: A Carolina Memoir.* New York: Free Press, 1983.

Marquart, Debra. *The Horizontal World: Growing Up in the Middle of Nowhere.* New York: Counterpoint, 2006.

Vossler, Ron. *Dakota Kraut: A Memoir.* Fargo, ND: Germans from Russia Heritage Collection, 2003.

Oral Histories

Bauer, Eva. Interview by Placid Gross. Richardton, ND (11 December 1997). Germans from Russia Heritage Collection, North Dakota State University, Fargo, ND.

Diede, Pauline (Neher). Interview by Bob Dambach. Hebron, ND (22 August 2002). Germans from Russia Heritage Collection, North Dakota State University, Fargo, ND.

Fischer, Emma. Interview by Ronald Vossler. Ashley, ND (2 July 1997). Germans from Russia Heritage Collection, North Dakota State University, Fargo, ND.

Hecker, Reinhardt. Interview by Michael M. Miller. Bismarck, ND (8 November 1993). Germans from Russia Heritage Collection, North Dakota State University, Fargo, ND.

Ketterling, Eva. Interview by Leland Elhard. Wishek, ND (2003). "Here by the Grace of God." St. Luke Lutheran Church. Germans from Russia Heritage Collection, North Dakota State University, Fargo, ND.

Larson, Annie (Roesch). Interview by Bob Dambach. Aberdeen, SD (13 August 1999). Germans from Russia Heritage Collection, North Dakota State University, Fargo, ND.

Risling, Barbara (Schneider). Interview by Bob Dambach. Scott, SK (19 July 2000). Germans from Russia Heritage Collection, North Dakota State University, Fargo, ND.

Saylor, Albert. Interview by Leland Elhard. Wishek, ND (2003). "Here by the Grace of God." St. Luke Lutheran Church. Germans from Russia Heritage Collection, North Dakota State University, Fargo, ND.

Symoniak, Anna. Interview by Leland Elhard. Wishek, ND (2003). "Here by the Grace of God." St. Luke Lutheran Church. Germans from Russia Heritage Collection, North Dakota State University, Fargo, ND.

Vossler, Ferdinand. Interview by Leland Elhard. Wishek, ND (2003). "Here by the Grace of God." St. Luke Lutheran Church. Germans from Russia Heritage Collection, North Dakota State University, Fargo, ND.

Welk, Thomas. Interview by Bob Dambach. Linton, ND (7 September 2001). Germans from Russia Heritage Collection, North Dakota State University, Fargo, ND.

SECONDARY SOURCES

"AHR Forum: History and Memory." *The American Historical Review* 102 (December 1997): 1371–412.

Allen, John L. "Landscape Change at the Confluence: From Lewis and Clark to the Present." *North Dakota History Journal* 69 (2002): 2-23.

Altschuler, Glen C. *All Shook Up: How Rock 'n' Roll Changed America.* Edited by David Hackett Fischer and James M. McPherson, Pivotal Movements in American History. New York: Oxford University Press, 2003.

"American Historical Society of Germans from Russia." http://www.ahsgr.org/organization.htm.

Anderson, Benedict. *Imagined Communities: Reflections on the Origin and Spread of Nationalism.* Revised ed. London: Verso, 1991.

Appleby, Joyce. "The Power of History." *American Historical Review* (February 1998): 1–14.

________, Lynn Hunt, and Margaret Jacob. *Telling the Truth about History.* New York: W.W. Norton and Company, 1994.

Arends, Shirley Fischer. *The Central Dakota Germans: Their History, Language, and Culture.* Washington, DC: Georgetown University Press, 1989.

Ariés, Philippe. *Centuries of Childhood: A Social History of Family Life.* New York: Random House, 1962.

Atkins, Annette. *We Grew up Together: Brothers and Sisters in Nineteenth-Century America.* Urbana: University of Illinois Press, 2001.

Baltensperger, Bradley H. "Agricultural Change among Great Plains Russian Germans." *Annals of the Association of American Geographers* 73 (March 1983): 75–88.

Barranti, Crystal C. Ramirez. "The Grandparent/Grandchild Relationship: Family Resource in an Era of Voluntary Bonds." *Family Relations* 34 (1985): 343–52.

Bridenthal, Renate. "Germans from Russia: The Political Network of a Double Diaspora." In *The Heimat Abroad: The Boundaries of Germanness.* Edited by Krista O'Donnell, Renate Bridenthal and Nancy Reagin, 187–218. Ann Arbor: The University of Michigan Press, 2008.

________. "Reaching Out: The Double Diaspora of Germans from Russia." Paper presented at the *Heimat* Abroad: The Boundaries of Germanness, New York University, 1999.

Burke, Peter. "History as Social Memory." In *Varieties of Cultural History.* Ithaca: Cornell University Press, 1997.

________. *What Is Cultural History?* Second ed. Cambridge: Polity, 2008.

Carr, E. H. *What Is History?* Second ed. *N.p.*: Palgrave, 2001.

Central Grasslands Research Extension Center. "Natural Resources of the Coteau Area of North Dakota." North Dakota State University. http://www.ag.ndsu.nodak.edu/streeter/99report/coteau.htm.

Chase Lake National Wildlife Refuge. "Wetlands." United States Fish and Wildlife. http://www.fws.gov/arrowwood/chaselake_nwr/wetlands.html.

Chinn, Sarah E. *Inventing Modern Adolescence: The Children of Immigrants in Turn-of-the-Century America.* New Brunswick: Rutgers University Press, 2009.

Chudacoff, Howard P. *Children at Play: An American History.* New York: New York University Press, 2007.

Clark, Jessica. "The German-Russian Christmas: Oral Histories from the Northern Plains." *Heritage Review* 37 (2007): 8–9, 30, and 44.

________. "Holocaust Education and Collective Memory in West Germany, 1945–1970." Master's thesis, North Dakota State University, 2004.

Clark, Jessica and Thomas D. Isern. "Germans from Russia: An Oral History." *American Review of Canadian Studies* 40 (March 2010): 71–85.

Connelly, Bridget. *Forgetting Ireland: Uncovering a Family's Secret Past.* St. Paul, MN: Borealis Books, 2003.

Conzen, Kathleen Neils. "Historical Approaches to the Study of Rural Ethnic Communities." In *Ethnicity on the Great Plains.* Edited by Frederick C. Luebke, 1–18. Lincoln: University of Nebraska Press, 1980.

________. "A Saga of Families." In *The Oxford History of the American West,* edited by Clyde A. Milner II, Carol A. O'Conner and Martha A. Sandweiss, 313–57. New York: Oxford University Press, 1994.

Coomber, James, and Sheldon Green. *Magnificent Churches on the Prairie: A Story of Immigrant Priests, Builders and Homesteaders.* Fargo, ND: Institute for Regional Studies, NDSU, 1996.

Cross, Gary. "Gendered Futures/Gendered Fantasies: Toys as Representations of Changing Childhoods." *The American Journal of Semiotics* 12 (1998): 289–310.

________. *Kids' Stuff: Toys and the Changing World of American Childhood.* Cambridge: Harvard University Press, 1997.

________. "Toys and Time: Playthings and Parents' Attitudes toward Change in Early 20th-Century America." *Time and Society* 7 (1998): 5–24.

Cunningham, Hugh. *Children and Childhood in Western Society Since 1500.* Edited by John Morrill and David Cannadine. 2nd ed. Studies in Modern History. London: Pearson Longman, 2005.

Danbom, David B. *Born in the Country: A History of Rural America.* Baltimore: The John Hopkins University Press, 1995.

________. *Going It Alone: Fargo Grapples with the Great Depression.* St. Paul, MN: Minnesota Historical Society Press, 2005.

Davidoff, Leonore, and Catherine Hall. *Family Fortunes: Men and Women of the English Middle Class 1780–1850.* Revised ed. London: Routledge, 2002.

Denham, Thomas, and Craig Smith. "The Influence of Grandparents on Grandchildren: A Review of the Literature and Resources." *Family Relations* 38 (1989): 345–50.

Dewald, Sandi. *Gazing Forward Glancing Back Remembering Always: Memories Retold and Relived by the Community of Streeter, North Dakota.* Fargo, ND: Germans from Russia Heritage Collection, North Dakota State University Libraries, 2000.

Dunaway, David K., and Willa K. Baum, eds. *Oral History: An Interdisciplinary Anthology.* New York: AltaMira Press, 1996.

Egerton, John. *Generations: An American Family.* Lexington: The University Press of Kentucky, 2003.

Elder, Glen H. Jr. *25th Anniversary Edition of Children of the Great Depression.* Boulder, CO: Westview Press, 1999.

Etulain, Richard W., ed. *Western Lives: A Biographical History of the American West.* Albuquerque: University of New Mexico, 2004.

Fass, Paula S. *Children of a New World: Society, Culture, and Globalization.* New York: New York University Press, 2000.

Fass, Paula S. and Mary Ann Mason, eds. *Childhood in America.* New York: New York University Press, 2000.

Frisch, Michael. *A Shared Authority: Essays on the Craft and Meaning of Oral and Public History.* New York: State University of New Press, 1990.

Fuller, Wayne E. *The Old Country School: The Story of Rural Education in the Middle West.* Chicago: University of Chicago Press, 1982.

________. "School District 37: Prairie Community." *The Western Historical Quarterly* (October 1981): 419–32.

Garceau-Hagen, Dee. "C.W.W.H. Roundtable Gender, Region, and Culture: New Voice and Directions." Roundtable at the *Western History Association*, Denver, CO, October 7, 2009.

Geertz, Clifford. "Thick Description: Toward an Interpretative Theory of Culture." In *The Interpretation of Cultures: Selected Essays*, 3–30. New York: Basic Books, Inc., Publishers, 1973.

"The Germans from Russia: Children of the Steppe, Children of the Prairie." United States: Prairie Public Broadcasting, Inc., 1999.

"Germans from Russia Heritage Society." http://www.grhs.org/.

"Germans from Russia Heritage Collection." North Dakota State University, www.ndsu.edu/grhc.

Gjerde, Jon. *The Minds of the West: Ethnocultural Evolution in the Rural Middle West, 1830–1917.* Chapel Hill: The University of North Carolina Press, 1999.

"Grasslands." United States Fish and Wildlife Service. http://www.fws.gov/arrowwood/chaselake_nwr/grasslands.html.

Greene, Roberta and Jirina S. Polivka. "The Meaning of Grandparents' Day Cards: An Analysis of the Intergenerational Network." *Family Relations* 34, no. 2 (1985): 221–25.

Halbwachs, Maurice. *The Collective Memory.* Translated by Francis J. Ditter Jr. and Vida Yazdi Ditter. New York: Harper Colophon Books, 1950.

________. *On Collective Memory.* Translated by Lewis A. Coser. Edited by Lewis A. Coser. Chicago: University of Chicago Press, 1992.

Hampsten, Elizabeth. *Settlers' Children: Growing up on the Great Plains.* Norman: University of Oklahoma Press, 1991.

Hanawalt, Barbara A. *Growing up in Medieval London: The Experience of Childhood in History.* New York: Oxford University Press, 1993.

Handlin, Oscar. *The Uprooted.* Second ed. Boston: Little, Brown and Company, 1973.

Handy-Marchello, Barbara. "Assimilation of German-Russians and Norwegian Immigrants." *Germans from Russia Heritage Society Heritage Review* 17 (May 1987): 29–32.

________. ""Extension Work among Black Sea Germans in Logan, McIntosh and Emmons Counties, 1933–1940." Master's thesis, North Dakota State University, 1988.

________. *Women of the Northern Plains: Gender and Settlement on the Homestead Frontier.* St. Paul, MN: Minnesota Historical Society Press, 2005.

Hareven, Tamara K. "The History of the Family and the Complexity of Social Change." *The American Historical Review* 96 (February 1991): 95–124.

Height, Joseph S. "*Christbaum* and *Christkindel.*" *North Dakota Historical Society of Germans from Russia Heritage Review* 16 (1976): 2–7.

________. *Homesteaders on the Steppe: Odyssey of a Pioneering People.* 2nd ed. Bismarck, ND: Germans from Russia Heritage Society, 1987.

________. *Memories of the Black Sea Germans from Russia: Odyssey of a Pioneering People.* Bismarck, ND: Germans from Russia Historical Society, 1979.

________. *Paradise on the Steppe: The Odyssey of a Pioneering People.* 5th ed. Bismarck, ND: Germans from Russia Historical Society, 1989.

Heywood, Colin. *A History of Childhood: Children and Childhood in the West from Medieval to Modern Times.* London: Polity Press, 2001.

Hildebrand, John. *Mapping the Farm: The Chronicle of a Farm.* New York: Alfred A. Knopf, 1995.

Holt, Marilyn Irvin. *Children of the Western Plains: The Nineteenth-Century Experience.* Chicago: Ivan R. Dee, 2003.

Hudson, Lois P. *Bones of Plenty.* Saint Paul, MN: Minnesota Historical Society Press, 1984.

________. *Reapers of the Dust: A Prairie Chronicle.* Minneapolis, MN: Minnesota Historical Society Press, 1984.

Hundley, Elaine (Helbling). "The Germans in Russia." *German-American Journal* 54 (2005): 13.

Iseminger, Gordon L. "Are We Germans, or Russians, or Americans?: The McIntosh County German-Russians During World War I." *North Dakota History: Journal of the Northern Plains* 59 (Spring 1992): 2–16.

________. "'Hunhuskers,' 'Red Cross Roosters,' and 'Uncle Sam Whiskers': The McIntosh County German-Russians in World War I." *Germans from Russia Heritage Society Heritage Review* 27 (December 1997): 11–20.

________. "The McIntosh County German-Russians: The First Fifty Years." *Journal of the American Historical Society of Germans from Russia* 11 (1988): 1–20.

Isern, Thomas D. *Dakota Circle: Excursions on the True Plains.* Fargo, ND: Institute for Regional Studies, North Dakota State University, 2000.

Jeffery, Jaclyn, and Glenace Edwall, eds. *Memory and History: Essays on Recalling and Interpreting Experience.* London: University Press of America, Inc., 1994.

Kinbacher, Kurt E. "Life in the Russian Bottoms: Community Building and Identity Transformation among Germans from Russia in Lincoln, Nebraska, 1876 to 1926." *Journal of American Ethnic History* (2007): 27–57.

King, Valerie, and Glen H. Elder Jr. "American Children View Their Grandparents: Linked Lives across Three Generations." *Journal of Marriage and the Family* 57 (1995): 165–78.

King, Valerie, and Glen H. Elder Jr. "The Legacy of Grandparenting: Childhood Experiences with Grandparents and Current Involvement with Grandchildren." *Journal of Marriage and the Family* 59 (November 1997): 848–59.

Klapper, Melissa R. *Small Strangers: The Experiences of Immigrant Children in America,* 1880–1925. Chicago: Ivan R. Dee, 2007.

Kloberdanz, Timothy J. "The Germans from Russia: A Viable Ethnic Group or Fading Phenomenon." *Journal of the American Historical Society of Germans from Russia* 27 (2004): 14–26.

________. "'The Tradition of Brauche': Facts and Fallacies." *Germans from Russia Heritage Society Heritage Review* 15 (September 1985): 3–5.

________. "'The Traditions We Share': Folklore in Four Cultures on the Northern Plains." *North Dakota History: Journal of the Northern Plains* 65 (1998): 19–31.

________. "*Volksdeutsche*, the Eastern European Germans: Hungry for Land, Hungry for a Home: North Dakota's 'Other Germans.'" In *Plains Folk: North Dakota's Ethnic History*, 117–81. Fargo, ND: Institute for Regional Studies at North Dakota State University, 1988.

Kroeger, Arthur. *Hard Passage: A Mennonite Family's Long Journey from Russia to Canada*. Edmonton, AB: The University of Alberta Press, 2007.

Lahlum, Lori Ann. "Growing up in Norwegian-American Communities: A Preliminary Study of Childhood, Adolescence, and Young Adulthood." In *Norwegian-American Essays 2008:* "Migration and Memory." Edited by Øyvind Tveitereid Gulliksen and Harry T. Cleven, 105–36. Oslo, Norway: NAHA, 2008.

Langer, Lawrence L. *Holocaust Testimonies: The Ruins of Memory.* New Haven: Yale University Press, 1991.

Layman, Lenore. "Reticence in Oral History Interviews." *Oral History Review* 36 (2009): 207–30.

Lindgren, H. Elaine. "Ethnic Women Homesteading on the Plains of North Dakota." *Journal of the American Historical Society of Germans from Russia* 13 (1990): 15–25.

________. *Land in Her Own Name: Women Homesteaders in North Dakota.* Forward by Elizabeth Jameson. Norman: University of Oklahoma Press, 1991.

Luebke, Frederick C. "Ethnic Group Settlements on the Great Plains." *Western History Quarterly* 8 (1977): 405–30.

________, ed. *Ethnicity on the Great Plains.* Lincoln: University of Nebraska Press, 1980.

Mause, Llyod de. *The History of Childhood.* New York: Psychohistory Press, 1974.

Mergen, Bernard. "The Discovery of Children's Play." *American Quarterly* 27 (October 1975): 399–420.

Miller, Michael M. "A Brief History of the Germans from Russia." Germans from Russia Heritage Collection, North Dakota State University, ndsu.edu/grhc/history_culture/history.

________. *Researching the Germans from Russia.* Fargo, ND: Institute for Regional Studies and North Dakota State University Library, 1987.

Mintz, Steven. *Huck's Raft: A History of American Childhood.* Cambridge: Belknap Press of Harvard University, 2004.

Napoleon 75th Diamond Jubilee Anniversary: 1884–1959. Napoleon, ND: Napoleon Diamond Jubilee Book Committee, 1959.

Napoleon Homestead, 1935–1939.

National Oceanic and Atmospheric Administration. "National Weather Service." United States Department of Commerce. weather.gov.

Neils, Kathleen. "A Saga of Families." In *The Oxford History of the America West*, edited by Clyde A. Milner II, Carol A. O'Conner, and Martha A. Sandweiss, 313–57. New York: Oxford University Press, 1994.

Nelson, Paula M. *After the West Was Won: Homesteaders and Town Builders in Western South Dakota, 1900–1917.* Iowa City, IA: University of Iowa Press, 1986.

________. *The Prairie Winnows Out Its Own: The West River Country of South Dakota.* Iowa City, IA: University of Iowa Press, 1996.

Neth, Mary. *Preserving the Family Farm: Women, Community, and the Foundations of Agribusiness in the Midwest, 1900–1940.* Baltimore: John Hopkins University Press, 1995.

"North Dakota: A Guide to the Northern Prairie State." In *Workers of the Federal Writers Project of the Works Progress Administration for the State of North Dakota.* Fargo, ND: Knight Printing Company, 1938.

Otto, John S. "Farming in Russia and North Dakota: One German-Russian Family's Experience." *North Dakota History: Journal of the Northern Plains* 55 (Spring 1988): 23--30.

________. "From the Russian Steppes to the North Dakota Prairies: The Agricultural Practices of a German-Russian Family." *Journal of the American Historical Society of Germans from Russia* 7 (1984): 37–44.

Pedeliski, Theodore B. "The German-Russian Ethnic Factor in William Langer's Campaigns, 1914–1940." *North Dakota History: Journal of the Northern Plains* 64 (Winter 1997): 2–20.

Pederson, Jane Marie. *Between Memory and Reality: Family and Community in Rural Wisconsin, 1870–1970.* Madison, WI: The University of Wisconsin Press, 1992.

Perks, Robert, and Alistair Thomson, eds. *The Oral History Reader.* London: Routledge, 2004.

Pollock, Linda A. *Forgotten Children: Parent-Child Relations from 1500–1900.* New York: Cambridge University Press, 1983.

Portelli, Alessandro. *The Order Has Been Carried Out: History, Memory, and Meaning of a Nazi Massacre in Rome.* New York: Palgrave, 2003.

________. *They Say in Harlan County: An Oral History.* Oxford University Press, 2011.

Prairies, 1978–1985.

Rader, Irene M. "The Experience of the German-Russian Pioneer Women." *Journal of the American Historical Society of Germans from Russia* 20 (1997): 31–7.

Rath, George. *The Black Sea Germans in the Dakotas.* Freeman, SD: Pine Hill Press, 1977.

________. "The Russian-Germans in the United States of North America, Preparations for Immigration." *Germans from Russia Heritage Society Heritage Review* 27 (September 1997): 9–22.

Riney-Kehrberg, Pamela. *Childhood on the Farm: Work, Play and the Coming of Age.* Lawrence: University of Kansas, 2005.

Rippley, LaVern J. "F.W. Sallet and the *Dakota Freie Presse.*" *North Dakota History: Journal of the Northern Plains* 59 (Fall 1992): 2–21.

Ritchie, Donald A. *Doing Oral History: A Practical Guide.* Oxford University Press, 2003.

Robinson, Elwyn B. *History of North Dakota.* Lincoln: University of Nebraska Press, 1966.

Roth, Jean. "Ellis Island: Gateway to America." *Journal of the American Historical Society of Germans from Russia* 22 (1999): 13–21.

Rozum, Molly P. *Grasslands Grown: Creating Place on the U.S. Northern Plains and Canadian Prairies.* Lincoln: University of Nebraska Press, 2021.

Salamon, Sonya. *Prairie Patrimony: Family, Farming & Community in the Midwest.* Chapel Hill: The University of North Carolina Press, 1992.

Sallet, Richard. *Russian-German Settlements in the United States.* Translated by LaVern Rippley and Armand Bauer. Fargo, ND: Institute for Regional Studies, 1974.

Sanders, Gregory F. and Debra W. Trygstad. "Stepgrandparents and Grandparents: The View from Young Adults." *Family Relations* 38 (January 1989): 71–75

________. "Strengths in the Grandparent-Grandchild Relationship." *Activities, Adaptation, and Aging* 17 (1993): 43–53.

Sanders, Gregory F., Joe Pittman, and James E. Montgomery. "Family Caregivers of Centenarians: Support for the Very Old." *The Journal of Applied Gerontology* 5 (December 1986): 113–25.

Schacter, Daniel L. *Searching for Memory: The Brain, the Mind, and the Past.* New York: Basic Books, 1996.

Schrems, Suzanne H. "Teaching School on the Western Frontier: Acceptable Occupation for Nineteenth-Century Women." *Montana: The Magazine of Western History* (Summer 1987): 54–63.

Schwartzman, Helen B. "The Anthropological Study of Children's Play." *Annual Review of Anthropology* 5 (1976): 289–328.

Schwieder, Dorothy Hubbard. *Growing Up with the Town: Family and Community on the Great Plains.* Iowa City, IA: University of Iowa Press, 2002.

Scott, Joan W. "Gender: A Useful Category of Analysis." *The American Historical Review* 91 (December 1986): 289–328.

Seabloom, Robert W. *The Mammals of North Dakota.* 2nd ed. Fargo, ND: North Dakota State University Press, 2020.

Sewell, William H., Jr. "Geertz, Cultural Systems, and History: From Synchrony to Transformation." *Representations* 59 Special Issue: The Fate of "Culture": Geertz and Beyond (Summer 1997): 1053–75.

Sherman, William C., Warren A. Henke, Theodore B. Pedeliski, Playford V. Thorson, Timothy J. Kloberdanz, and Robert P. Wilkins. *Plains Folk: North Dakota's Ethnic History.* Fargo, ND: Institute for Regional Studies at North Dakota State University, 1988.

Shopes, Linda. "What Is Oral History?: History Matters: The Us Survey on the Web." historymatters.gmu.edu.

Shorter, Edward. *The Making of the Modern Family.* London: Collins, 1976.

"The Statue of Liberty—Ellis Island Foundation, Inc." ellisisland.org.

Stegner, Wallace. *Wolf Willow: A History, a Story, and a Memory of the Last Plains Frontier.* New York: Penguin Books, 2000.

Stone, Lawrence. *Family, Sex and Marriage in England* 1500–1800. New York: Harper and Row, 1977.

Stumpp, Karl. *The Emigration from Germany to Russia in the Years 1763 to 1862.* Vol. 1. Lincoln, NE: American Historical Society of Germans from Russia, 1973.

________. *The Emigration from Germany to Russia in the Years 1763 to 1862.* Vol. 2. Lincoln, NE: American Historical Society of Germans from Russia, 1973.

Sutton-Smith, Brian, ed. *Play and Learning.* New York: Gardner Press, 1979.

Teeuwen, Randall C. "The Immigration and Early Education of Germans from Russia." *Journal of the American Historical Society of Germans from Russia* 8 (1985): 52–57.

Thompson, Paul. *The Voice of the Past: Oral History.* Oxford: Oxford University Press, 2000.

Thomson, Alistair. "Fifty Years On: An International Perspective on Oral History." *The Journal of American History* 85 (September 1998): 581–95.

Turner, Victor. “Liminality and Communitas.” In *The Ritual Process: Structure and Anti-Structure*, 94–130. Ithaca, NY: Cornell University Press, 1969.

Vossler, Ronald. “Prairie *Spass*: The Folk Humor of North Dakota's Germans from Russia.” *North Dakota History: Journal of the Northern Plains* 66 (Summer/Fall 1999): 50–60.

United States Geological Survey. “Climate of North Dakota.” United States Department of Interior. npwrc.usgs.gov/resource/habitat/climate/precip.htm.

________. “Native Wildflowers of the North Dakota Grasslands: Common Pincushion Cactus (*Coryphantha Vivipara*).” United States Department of Interior. npwrc.usgs.gov/resource/plant/wildflwr/species/coryviv.htm.

________. “Prairie Wildflowers and Grasses of North Dakota: Spring Flowers.” United States Department of Interior. npwrc.usgs.gov/resource/plant/pwgofnd/prairc.htm.

Wagner, Jonathan F. “In Pursuit of the *Gesundheit*: Changing Health Practices among North Dakota's German-Russians, 1890–1930.” *North Dakota History: Journal of the Northern Plains* 58 (Summer 1991): 2–15.

Webb, Walter Prescott. *The Great Plains*. Boston: Ginn & Company, 1931.

West, Elliott. *The Contested Plains: Indians, Goldseekers, and the Rush to Colorado*. Lawrence: University of Kansas Press, 1989.

________. *Growing Up in Twentieth-Century American: A History and Reference Guide*. West Port, Connecticut: Greenwood Press, 1996.

________. *Growing Up with the Country: Childhood on the Far Western Frontier*. Albuquerque: University of New Mexico Press, 1989.

West, Elliott and Paula Petrick, eds. *Small Worlds: Children and Adolescents in America*, 1850–1950. Lawrence: University of Kansas Press, 1992.

White, Clinton O. “Education among German Catholic Settlers in Saskatchewan: A Reinterpretation.” *Canadian Ethnic Studies* 16, no. 1 (1984): 78–95.

________. “Maintaining Anglo-Celtic Cultural Hegemony in Saskatchewan: Rev. E. H. Oliver, Provincial Educational Policy, and the German Catholics.” *Historical Studies in Education* 6, no. 2 (1994): 253-79.

________. “The Politics of Elementary Schools in a German American Roman Catholic Settlement in Canada's Province of Saskatchewan, 1903–1925.” *Great Plains Center for Great Plains Research: A Journal of Natural and Social Sciences* 7 (Fall 1997): 251–72.

________. “Pre-World War I Saskatchewan German Catholic Thought Concerning the Perpetuation of Their Language and Religion.” *Canadian Ethnic Studies* 26, no. 2 (1994).

White, Richard. *Remembering Ahanagran: Storytelling in a Family's Past*. New York: Hill and Wang, 1998.

Wishek News, 1935–1939.

Wishek, North Dakota, 1898–1998. Wishek, ND: Centennial Book Committee, 1998.

Wrobel, David M. *Promised Lands: Promotion, Memory, and the Creation of the American West*. Lawrence University of Kansas Press, 2002.
Zelizer, Viviana. *Pricing the Priceless Child: The Changing Social Value of Children*. Princeton: Princeton University Press, 1985.

APPENDIX A. INTERVIEW QUESTIONS

The founders of the Dakota Memories Oral History Project (DMOHP) used Paul Thompson's *The Voice of the Past* to revise the existing oral history questions of the Germans from Russia Heritage Collection (GRHC). Ann Braatan, Dale Davis, Shirley (Welk) Fredricks, Margaret Freeman, Jay Gage, LaVerne (Ruff) Graff, Placid Gross, Betty Maier, Chris Maier, Michael M. Miller, Cheryl Schumacher, Roland Wagner, Catherine (Glatt) Wald, and Michele Ziebarth wrote the original questions. In 2005, Jessica Holkup, Tom Isern, Danielle Stuckle, and I revised these questions. These new questions became the interview guide for DMOHP oral history teams to use during interviews.

Introductory Remarks

Today is ________________ [date]. I am ________________ [name of interviewer]. It is a pleasure to conduct this interview for the Dakota Memories Oral History Project in ________________ [city, state].

Introductory Questions

Please state your full name (including maiden name).
When and where were you born?
Have you ever heard an interesting story about your birth? If so, please share.
Was a midwife or doctor present? Please elaborate.
Please share some of your earliest memories.

Environment

What was it like growing up on the plains?
What did you find beautiful and/or enjoyable about growing up on the plains?
What did you find bleak and/or unpleasant about growing up on the plains?
What were winters like?
Please describe a typical fall.
What were summers like?
Please describe a typical spring.
What was your favorite season? Why?

Family

GRANDPARENTS

Please tell us about your grandparents.
What were their names?
What type of person was your grandmother?
What type of person was your grandfather?
Did your grandparents come over from Russia? If so, where did they come from, and where did the settle?

Have you heard any stories about their lives in Russia, their voyage to the US, or their journey to across the US to the Great Plains? If so, please share.
Where did they live when you were growing up?
What did they do for a living when you were growing up?
What did you do when you visited your grandparents?
How important was religion in your grandparents' lives? Explain.
How many children did they have?
Were there any other relatives of the older generations (great uncles and aunts) present in your childhood? If so, please tell us about them and their role in your life.
Is there anything else you would like to add about your grandparents or others from their generation?

EXTENDED FAMILY (*UNCLES, AUNTS, COUSINS)*

Please tell me about your uncles and aunts.
Where did they live when you were growing up?
What was it like visiting them?
Who was your favorite aunt/uncle? Why?
Who was your favorite cousin? What was s/he like?
Describe your family gatherings.
How did the families stay in touch with each other?
Is there anything else you would like to share about your extended family members?

PARENTS

Father

Please tell us about your father
Where and when was he born?
How did your father express his emotions, such as fear, love, and anger?
Describe your relationship with your father.
What was his profession?
How did your father cope during hard times?
When, where, and how did your father die?

Mother

Please tell us about your mother.
Where and when was she born?
How did she express her emotions, such as fear, love, and anger?
Describe your relationship with your mother.
What was her profession?
How did she cope during hard times?
When, where, and how did she die?
How many children did they have?

Who was the strict parent— your mother or father?
How would they discipline the children?
Who influenced your life more—your mother or your father? How so?
How did your parents meet?
How long were they married?
Please describe their marriage.
What types of things did your parents do for pastimes?
Were your parents involved in politics when you were a child? If so, please tell us about it.
Do you have anything else you would like to add about your parents?

SIBLINGS

Please tell me the birth order of your siblings.
Tell me about your brothers and sisters.
Where and when were they born?
Was anyone present at their births?
Please share any interesting stories about their births.
Growing up, how did the children get along with each other?
Was there any competition in your family?
What type of mischief did the children get into?
Which brother/sister were you closest to? Please describe your relationship.
What was it like when a new baby came into the home?
What was it like when your brothers and sisters left home?
Did your siblings have nicknames? If so, please share them with us. What are the stories behind the nicknames?
Is there anything else you would like to add about your brothers and sisters?

ONLY CHILD

Please tell us what it was like growing up an only child?
Why didn't your parents have more children?
Did you want a brother or sister?
Is there anything else you would like to add about growing up as an only child?

Play

What type of activities did you partake in for fun?
What games/activities did you play inside?
What games/activities did you play outside?
Where did you learn how to play these games/activities?
How did the weather influence the games/activities you played?
Do you remember playing any games/activities that your parents/grandparents learned to play while they were living in Russia?

What type of toys did you play with as a child? Were they store-bought or homemade? Do you still have any of your childhood toys? If so, can we see them?
Who did you play with?
Tell us about your parents' involvement in pastimes.
Where was your favorite place to play? Why?
Did you ever play with the farm animals? How so?
Did you go hunting as a child? If so, please tell us about it.
Did you go trapping as a child? If so, please tell us about it.
Did you go fishing as a child? If so, please tell us about it.
Did you have any favorite rhymes or sayings? Can you please share them with us?
Please tell me about any superstitions or scary stories you grew up with?
What were your favorite songs as a child? Will you sing one of them?
Did anyone in your family sing or play musical instruments? Please tell us more about this.
Were you encouraged to play an instrument? Why?
Did you take lessons or were you self-taught? Can you read music?
Is there anything else you would like to share with us about your childhood pastimes?

Grade School

What did you want to be when you grew up?
Where did you go to school? Was this a one-room schoolhouse?
How did you get to school? How long did it take you to get to school?
What were your schools like? (classrooms, population, teachers, supplies, et cetera)
What was the composition of the school population? (gender and ethnicity)
Did your teachers treat children from different social backgrounds or ethnicity differently? Explain.
Did children treat other children from different social or ethnic backgrounds differently? How so?
Please describe the attendance situation in your school.
At times, did you have to stay home to help with work?
What is your favorite memory about going to school?
What subjects were you taught?
Were you a good student? How so?
Did you have homework? Please describe it.
Did you parents and/or siblings help you with your homework? Explain.
What books did you read at school? Home?
Did you get into trouble at school? If yes, how were you disciplined?

What language did you speak in school? Could you speak German? Was it hard learning English? During recess or noon hour, did children speak German or English? Did you help teach your parents English?

What did you take for lunch? In addition, how did you take your lunch to school?

Did you change country schools or go to a town school? If so, what were some of the differences between the schools? Which school did you like better – the town school or country school? Why?

Do you remember your teachers? How often did you get a new teacher? Were the teachers male or female, young or old? Please tell me about them.

What social activities were allowed during "recess" and "noon-hour"?

Did your schools have parties to celebrate the holidays, such as Halloween, Christmas, and Valentine's Day? Please describe these occasions.

Were you encouraged or discouraged from going to high school? Why?

How important was getting an education to your family?

Chores

Please tell us about the chores you did as a child.

Who did what around the house: the cooking/cleaning/washing/et cetera?

Did you help around the house? How so?

Did you help care for your family's garden? How so?

What roles did your brothers and sisters have around the house and garden?

Did your father help with the household tasks? If so, how?

How would your family do laundry when you were growing up? Where did you hang the clothes in the winter? How often did your family do the laundry? How would you iron clothes?

What were some of your outside chores?

Did your mother work out in the fields or outside the home when you were growing up? What did she do? How often?

Which chores did you enjoy, and which chores did you dislike?

Were their differences between winter chores and summer chores? If so, please tell us what the differences were.

Were you ever hired out to work for a neighbor or relative? If so, please tell us about this experience.

Daily Life

FAMILY HOME

Please tell me about the house you grew up in? How many rooms did it have? Where did you all sleep in the house? What was the furniture like? How long did your family reside in this house? Who lived in your home when you were growing up?

If your family moved, where, when, and why did they move?

How did your family keep the house warm in the winters? How did your family keep the house cool in the summers?
How often would your family entertain guests in your home? Please tell us what it was like when your parents entertained guests.
What language did you speak at home? Were you allowed to speak English at home? How did your family feel about speaking English?
Where did you go to the bathroom? Please describe this building (location, toiletries, et cetera).
Please tell us about bathes. How often would you take a bath? Did you make your own soap? How did you heat the water? How often did you change the water?
How often were you supposed to brush your teeth? How often did you brush your teeth?
Did your mother have a garden? Please tell us about it.

FOOD

Who usually prepared the meals in your family?
Where were the meals usually prepared? What about during the summer months?
Where did you store food to keep it cool during the summer?
Did boys help in the kitchen? If so, please tell us more about their roles.
How would your family have their meals? Would you always sit at a table together? Were there assigned seats? Were you allowed to talk?
What types of food did your family usually eat?
What were some of your favorite meals?
What were some of your least favorite meals?
What were the traditional German dishes that you had growing up?
Did your family make ice cream? How would you do this?
Did your family do their own canning, butchering, and/or sausage smoking? If so, please tell us more about the processes.
What was your family's attitude toward alcohol? How often did people drink alcohol when you were growing up?
Did your family make their own root beer, beer, wine, or liquor? If so, please tell us more about the processes.

CLOTHING

What types of clothes were common when you were growing up?
Were your clothes store-bought or homemade?
If you had store-bought clothes, were would your parents get your clothes? Stores or catalogs? How often did you get new cloths?
If you had homemade clothes, who made them for you? Where did he/she get the material? Did he/she ever reuse old clothes to make new clothes?
Did you ever have to wear hand-me-down clothes?

Did anyone in your family sew, embroider, or quilt? Please tell us a little about this.

How often did you get new shoes? How many pair of shoes did you have as a child?

TECHNOLOGY

When did your family get electricity? How did this change your life?

When did your family get its first car or truck? How did this change your family's life?

When did your family get a telephone? How did this change your life?

How important was the radio in your family? What were some radio programs that your family listened to often?

Farm Life

Describe the homestead/farmstead you grew up on.

Where was the closest town?

Did you ever experience isolation? If so, please tell us about this experience.

How did your family gain possession of the land – did they inherit it, buy it, et cetera?

Was your family "grain farmers," "ranchers," dairy farmers, or diversified?

What buildings were on your parent's farm? What were they used for?

What type of animals did your family have? What did you use them for?

Did your family butcher these animals? Please tell me about it.

Describe the roads to your farm.

What type of farming equipment did your family have?

Please describe the fields and crops on your family's farm.

Do you remember your neighbors? Who were they? How close did they live to your farm? Did they ever look after the children, while your parents were away? If so, did they ever discipline you?

Was "neighboring" a common practice when you were growing up? If so, please explain.

What were the farm kids' feelings toward the town kids?

What was your family's attitude toward town people?

What were the positive aspects of growing up on a farm? How about the negative ones?

What social classes existed when you were growing up? What social class did your family belong to?

What ethnic groups were included in your community?

What was your family's attitude toward different ethnic groups, such as Norwegians or Scandinavians?

Were there places that your parents would not let you go or children they wouldn't let you socialize with? If yes, why?

Town Life

Describe the town you grew up in.
Who founded the town?
What was the population when you were growing up?
What is the present population?
What businesses were there in town?
How many churches were in your town?
What denominations were they?
Who were your neighbors? How close did they live to your family? Did they ever look after you? If you stepped out of line, would they discipline you?
What social classes existed when you were growing up?
What social class did your family belong to?
Were there other ethnic groups in the town you grew up in? If so, what ethnic groups?
What was your family's attitude toward different ethnic groups, such as Norwegians or Scandinavians?
Which ethnic groups were rich, and which were poor?
What were the town kids' feelings toward the farm kids?
What about the relationship between town adults and farm adults?
Describe a typical Saturday night in your town?
What were the positive aspects of growing up in a small town? How about the negative ones?
Describe some of the "characters" you remember in your town.
Were there places your parents would not let you go to or children they would not let you socialize with? If yes, why?

Religion

How important was religion during your childhood?
How often did your family attend church?
What denominations were you raised?
What church did your family attend?
Please tell us about the seating arrangements.
What were some of the religious activities that you participated in while you were growing up?
Were you baptized and/or confirmed? If so, please tell me about it.
Did your family have a celebration after your confirmation? If so, please tell us about it.
Were you able to question religious teachings?
What did your parents restrict because of religion?
Was Sunday dinner or supper different from other days of the week? If so, how?
What was the typical meal for a church celebration?

Where was the nearest church?
Was the German language used in church worships and prayers?
How did your family respond and react toward death?
Please describe a typical funeral.
How did your family view other faiths?

Holidays and Celebrations

What was your favorite holiday? Please describe it.
Did your family celebrate Christmas Eve and Christmas? If so, please describe the location, people, food, gifts, and games.
Did your family celebrate Easter? If so, please describe the location, people, food, gifts, and games.
Did your family celebrate Fourth of July? If so, please describe the location, people, food, gifts, and games.
Did your family celebrate birthdays? If so, please describe the location, people, food, gifts, and games.
Did your family celebrate names-day of your saint? If so, please describe the location, people, food, gifts, and entertainment.
Did your family participate in sausage parties, communal butchering, or quilting parties? If so, please tell me about them.
Please tell us about any other holidays that your family partook in during your childhood, such as Halloween or New Years.

Youth Activities and Dating

What types of pastime activities did you participate in during your teenage years?
Were you involved in any sports and/or clubs? If so, please elaborate.
What were some of your favorite radio programs?
Did you go to the movies? If so, please share your memories. What type of movies did you see? How often? Who went with you?
How old were you when you started dating?
How are the customs of courtship different than they are today?
Describe a typical date.
During your young adult years, did you ever attend parties? Describe this occasion.
Did you go dancing? If so, describe the occasion.
Who attended these dances— young people, older couples, and/or families?
What day were these dances typically held on?
What was the attitude of the older generation towards dance halls?
Was there a community-meeting place for people to socialize, whether in town or someone's farm or school?
How and what did you learn about married life? From parents?

Do you remember your parent's attitudes toward sex? Explain.
How would they have reacted to premarital sex?
Did they ever offer you advice on sex or relationships?
Whom did you go to if you needed that kind of advice?

High School

Please tell us about your experiences in high school.
How old were you when you started high school?
Where did you attend high school?
Did you have to board in town? Please tell us about this experience.
Please tell us about the transition from grade school to high school. Was it hard? Did a lot change?
What classes did you take?
What was the school like—class size, teacher to student ration, supplies, rooms, et cetera?
What was the composition of the school population—gender, ethnicity, et cetera?
Did teachers treat children different if they belonged to a low social class or to a minority group?
What was the attendance like in high school?
What is your favorite memory from your high school years?
Did you participate in any school activities—sports, clubs, et cetera? If so, please tell us more.
What was the coursework like in high school?
How would a student be discipline, if he/she stepped out of line?
Did your high school have pep rallies and dances? If so, please tell us a little about this.
Were you encouraged or discouraged from attending college or a normal school?
Is there anything else about high school that you would like to share with us today?

Health Care and Funerals

Please tell us about the health care that was available when you were growing up.
Was anyone in your family seriously ill or injured during your childhood?
What kind of help did they get?
Do you recall any home remedies or cures? Such as?
Do you remember hearing about births at home?
What do you remember about midwives?
Where was the closest doctor's office?
During your childhood, did you ever visit a dentist? Please elaborate on this.

During your childhood, did you ever visit an eye doctor? Please elaborate on this.

Did anyone in your family pass away during your childhood? If so, how?

Did you have a funeral for him/her? Will you describe it?

Is there anything else you would like to add about health care and funerals?

Historical Events

THE DIRTY THIRTIES

Please describe the Dirty Thirties.

How did the Great Depression affect your family?

Did you experience any dust storms? If so, please tell us more about this.

Did you experience any problems with grasshoppers, jackrabbits, or other animal pests during these years? If so, please share your memories with us.

How did your parents keep food on the table during the Dirty Thirties?

Do you remember the WPA? If so, please tell us about it.

Do you remember the CCCs? If so, please tell us about them.

Is there anything you would like to add about the Dirty Thirties?

WORLD WAR II

Please tell us how the outbreak of World War II changed your life.

How did rationing affect your family?

Did speaking German affect your relationship with others in school, town, or church?

Where were you when Pearl Harbor was attacked?

Was anyone in your family drafted? Alternatively, did anyone in your family volunteer to serve in the war? If so, please tell us more about it.

Did you serve in World War II? If so, where did you serve? Please tell us about your experiences.

KOREAN WAR

How did outbreak of the Korean War affect your family?

VIETNAM WAR

How did the outbreak of the Vietnam War affect your family?

WORLD TRADE CENTER ATTACK

How were you affected by the attack on the World Trade Center?

General Childhood Memories

What was your most stressful childhood experience?

What is your happiest childhood memory?

What was the most adventurous thing you did during your childhood?

Is there anything else about your childhood memories you would like to add?

Higher Education

Please tell me about your college experiences.
What did your parents think about you going to college?
Who funded your education?
Where did you go?
What degree (if any) did you earn? How? Why?
What is your favorite college memory?
What is your most difficult memory?
While you were at school, did you have any part-time jobs?
Is there anything else you would like to add about your college or higher education experiences?

Weddings and Traditions

How did you meet your spouse?
Was there any "arrangement" or "arranging" done by either family?
What was your parents' attitude toward your spouse?
When and where did you get married?
Will you tell me about the wedding and reception?
Were there money dances for the bride and groom? Explain.
What foods were served at weddings?
Anything unusual about your courtship and/or wedding?
Did have a honeymoon? If so, please tell me about it.
Is there anything else regarding wedding traditions that you would like to add?

Concluding Insights and Reflections

Why do you think it is important to tell your life story?
Are there other thoughts or observations you would like to share?
Is there anyone else you would like to suggest for us to interview?

APPENDIX B. SPECIAL TECHNIQUES

The special techniques employed by the Dakota Memories Oral History Project's (DMOHP) oral historians bring forth richer recollections of the past. Through the special techniques, the oral historians break past the barriers of collective memory and into the richness of individual memories, revealing the complexities of the new German-Russian story.

Yearly Round

This seasonal approach should not be used too early in the interview process. To employ this technique, determine a historical period. Then ask the narrator to identify the beginning of the average year. Prompt the narrator to proceed with storytelling about the events of the year (a chronological walkthrough).

Memory Map

This artistic memory trigger requires supplies—a sketchpad and pens. Do not use pencils, as they fad overtime. Ask the narrator to sketch a location near and dear to him or her. Encourage the narrator to describe the location, as he or she is sketching it. Be sure to keep the map, as the narrator just created an archival document.

Cemetery/Landmark Walk

This technique involves a walkthrough of physical areas important to the narrator's story. These walks should always be recorded via video footage and digital images. They are truly a living history.

Photo Album Tour

This visual stimulus involves the oral historian and narrator using images and documents to recreate the past. The narrator typically has no problems sharing stories about the images and documents under review.

Cooking/Music Demonstration

By demonstrating their talents, narrators share something they learned as a youngster, as well as stories about acquiring their skills. In between sampling food and music, ask the narrator for copies of recipes or sheet music, as well as stories about the past.

APPENDIX C. BIOGRAPHY FORM

Dakota Memories Oral History Project
Sponsored by the Germans from Russia Heritage Collection
Biography Form

Date: ____________________

Your Biography Information

Name: ____________________

Maiden Name (if applicable): ____________________

Mailing Address:
Unavailable*

Street Address:
Unavailable*

Phone Number: Unavailable*

E-Mail Address: Unavailable*

Birth Date: ____________________

Birth Place: ____________________

Occupation(s) (list all occupations chronologically):

Religion: ____________________

Schooling (list all schools attended, the dates of attendance, and note the level of education you have completed):

Clubs and Organizations (note dates of memberships and offices held):

Your Mother's Biography Information

Name:

Maiden Name (if applicable):

Birth Date:

Birth Place:

Occupation(s):

Date of Death:

Place of Death and Burial:

Her ancestor's village in Russia (Ukraine) or Bessarabia:

Her parents' names (include maiden name):

Her siblings' names (include maiden names):

Your Father's Biography Information

Name: ______________________________

Birth Date: ______________________________

Birth Place: ______________________________

Occupation(s): ______________________________

Date of Death: ______________________________

Place of Death and Burial: ______________________________

His ancestor's village in Russia (Ukraine) or Bessarabia: ______________________________

His parents' names (include maiden name): ______________________________

His siblings' names (include maiden names):

Your Brothers & Sisters' Biography Information

Names (include birthdates and locations, if known):

Your Spouse's Biography Information

Name: ______________________________

Maiden Name (if applicable): ______________________________

Birth Date: ______________________________

Birth Place: ______________________________

Wedding Date: ______________________________ **Wedding Place:** ______________________________

Occupation(s): ______________________________

Date of Death: ______________________________

Place of Death and Burial: ______________________________

Your Children's Biography Information

Names (include birthdates and locations, if known):

Family Histories

Does your family have their history published? If so, what is the name of the publication? How can we secure a copy of this publication?

***For unavailable information, contact Jessica Clark, Project Coordinator at (701) 231-8419 or Jessica.Clark@ndsu.edu.**

APPENDIX D. TOPIC CHECKLIST

GERMANS FROM RUSSIA
GRHC
HERITAGE COLLECTION
NDSU

Dakota Memories Oral History Project

Interview Topic Checklist

This will determine the scope of your interview. Please check all topics that you would like to discuss. The pre-checked boxes are standard topics that we address in every interview.

Name:______________________________

- ☐ **Introductory Remarks & Questions**
- ☐ **New Oral History Techniques**
- ☐ *The Yearly Round*
- ☐ *The Memory Map*
- ☐ *The Cemetery Walk*
- ☐ *The Landmark Walk*
- ☐ *The Family Album*
- ☐ **Environment**
- ☐ **Family**
- ☐ **Play**
- ☐ **Grade School**
- ☐ **Chores**
- ☐ **Daily Life**
- ☐ **Farm & Community Life**
- ☐ **Town & Community Life**
- ☐ **Religion**
- ☐ **Holidays & Celebrations**
- ☐ **Youth Activities & Dating**
- ☐ **High School**
- ☐ **Health Care & Funerals**
- ☐ **Historical Events**
- ☐ **Childhood Memories Conclusions**
- ☐ **Higher Education**
- ☐ **Weddings & Traditions**
- ☐ **Concluding Insights & Reflections**

APPENDIX E. INFORMED CONSENT

GERMANS FROM RUSSIA
GRHC
HERITAGE COLLECTION

Dakota Memories Oral History Project

Informed Consent Form Part I

I, ***Jessica Clark***, am a doctoral student in the History Department at North Dakota State University and Project Coordinator for the Dakota Memories Oral History Project. This project consists of conducting interviews, in order to preserve the history of Germans from Russia on the Northern Plains. The interviews will be made available to the public, as they will be videotaped, tape recorded, transcribed, and/or placed on the Germans from Russia Heritage Collection's webpage. Furthermore, I plan to use these interviews in my dissertation, which will focus on growing up German from Russia on the Northern Plains.

We, the ***Germans from Russia Heritage Collection staff***, invite you to participate in this research study. Your participation is voluntary, and you may withdraw from the project at anytime. If you decide to participate in this, please keep the first page of this packet for your information. Each interview will take approximately one or more hours, depending on your availability, comfort, and memories.

NOTE: Your identity will be revealed, as the project is designed to preserve your history.

If you have any questions please contact me, ***Jessica Clark***, at **(701) 231-8419** or Jessica.Clark@ndsu.edu. If for some reason you are unable to reach me, please contact either Dr. Tom Isern at **(701) 799-2942** or isern@plainsfolk.com, or Michael Miller at **(701) 231-5564** or Michael.Miller@ndsu.edu. If you have any questions about the rights of participants in research, or to report a problem, you should contact the NDSU IRB Office at **(701) 231-8908**.

We, the ***Germans from Russia Heritage Collection staff***, would like to thank you for your participation. Furthermore, we promise to supply you with a DVD copy of your oral history (interview).

Sincerely,

Jessica Clark,
Project Coordinator

(Interviewee's copy)

1

GERMANS FROM RUSSIA
GRHC
HERITAGE COLLECTION

Dakota Memories Oral History Project

Informed Consent Form Part II

I, ***Jessica Clark***, am a doctoral student in the History Department at North Dakota State University and Project Coordinator for the Dakota Memories Oral History Project. This project consists of conducting interviews, in order to preserve the history of Germans from Russia on the Northern Plains. The interviews will be made available to the public, as they will be videotaped, tape recorded, transcribed, and/or placed on the Germans from Russia Heritage Collection's webpage. Furthermore, I plan to use these interviews in my dissertation, which will focus on growing up German from Russia on the Northern Plains.

We, the ***Germans from Russia Heritage Collection staff***, invite you to participate in this research study. Your participation is voluntary, and you may withdraw from the project at anytime. If you decide to participate in this, please keep the first page of this packet for your information. Each interview will take approximately one or more hours, depending on your availability, comfort, and memories.

NOTE: Your identity will be revealed, as the project is designed to preserve your history.

If you have any questions please contact me, ***Jessica Clark***, at **(701) 231-8419** or Jessica.Clark@ndsu.edu. If for some reason you are unable to reach me, please contact either Dr. Tom Isern at **(701) 799-2942** or isern@plainsfolk.com, or Michael Miller at **(701) 231-5564** or Michael.Miller@ndsu.edu. If you have any questions about the rights of participants in research, or to report a problem, you should contact the NDSU IRB Office at **(701) 231-8908**.

We, the ***Germans from Russia Heritage Collection staff***, would like to thank you for your participation. Furthermore, we promise to supply you with a DVD copy of your oral history (interview).

Sincerely,

Jessica Clark,
Project Coordinator

GERMANS FROM RUSSIA
GRHC
HERITAGE COLLECTION

Dakota Memories Oral History Project

Informed Consent Form Part III

I am aware of all the rights and responsibilities involved with this interview. Furthermore, I give the ***Germans from Russia Heritage Collection, North Dakota State University Libraries*** ownership of the interview recorded on ____________________ (date) at _________________________ (place), including copyright. This includes worldwide rights of broadcast and distribution on any existing media including cassette, videotape, print, or electronic forms, such as CD-ROM and the Internet/Web, or on any media developed in the future. I warrant that all materials I furnish and use are my own original materials or are materials, which I have full authority to use for such purposes.

Narrator:

____
Print Name

______________________________________ _______________
Signature Date

Interviewer:

____
Print Name

______________________________________ _______________
Signature Date

Dakota Memories Oral History Project
Germans from Russia Heritage Collection
The Libraries
NDSU Dept #2080
PO Box 6050
Fargo, ND 58108-6050

INDEX

Page numbers in **bold** type indicate photographs or illustrations.

ABOUT THE AUTHOR

Jessica "Jess" Clark is a historian dedicated to uncovering and sharing the lived experiences of everyday people. Her research centers on community memory, cultural traditions, and the ways history shapes individual and collective identity.

She holds a PhD in History from North Dakota State University and a Post-Doctoral Graduate Certificate in Higher Education Leadership from the University of Wyoming. Over the course of her academic career, Jess has combined rigorous scholarship with a talent for accessible storytelling, bridging the gap between academic research and public audiences.

A first-generation college student, Jess has always believed in the power of education and historical understanding to transform lives and strengthen communities. She has presented her work widely and published on topics ranging from regional histories to cultural resilience.

Jess currently serves as Provost and Vice President of Academic Affairs at Bellevue College, Washington, where she continues to champion scholarly inquiry, student success, and inclusive learning environments.

ABOUT THE PRESS

North Dakota State University Press (NDSU Press) exists to stimulate and coordinate interdisciplinary regional scholarship. These regions include the Red River Valley, the state of North Dakota, the plains of North America (comprising both the Great Plains of the United States and the prairies of Canada), and comparable regions of other continents. We publish peer reviewed regional scholarship shaped by national and international events and comparative studies. Neither topic nor discipline limits the scope of NDSU Press publications. We consider manuscripts in any field of learning. We define our scope, however, by a regional focus in accord with the press's mission. Generally, works published by NDSU Press address regional life directly as the subject of study. Such works contribute to scholarly knowledge of region (that is, discovery of new knowledge) or to public consciousness of region (that is, dissemination of information or interpretation of regional experience). Where regions abroad are treated, either for comparison or because of ties to those North American regions of primary concern to the press, the linkages are made plain.

Since 1950, NDSU Press has published substantial trade books, but the line of publications is not limited to that genre. We also publish textbooks (at any level), reference books, anthologies, reprints, papers, proceedings, and monographs. The press also considers works of poetry or fiction, provided they are established regional classics, or they promise to assume landmark or reference status for the region. We select biographical or autobiographical works carefully for their prospective contribution to regional knowledge and culture. All publications, in whatever genre, are of such quality and substance as to embellish the imprint of NDSU Press.

We changed our imprint to North Dakota State University Press in January 2016. Prior to that, and since 1950, we published as the Institute for Regional Studies Press. We continue to operate under the umbrella of the Institute for Regional Studies, located at North Dakota State University.